T0257796

Developments in Mobile Robotics

Developments in Mobile Robotics

Edited by **Jared Kroff**

LANRYE
INTERNATIONAL

New Jersey

Published by Clanrye International,
55 Van Reypen Street,
Jersey City, NJ 07306, USA
www.clanryeinternational.com

Developments in Mobile Robotics
Edited by Jared Kroff

International Standard Book Number: 978-1-63240-141-0 (Hardback)

Contents

Preface

This book aims to highlight the current researches and provides a platform to further the scope of innovations in this area. This book is a product of the combined efforts of many researchers and scientists, after going through thorough studies and analysis from different parts of the world. The objective of this book is to provide the readers with the latest information of the field.

Numerous researches are being conducted in the field of robotics and mobile robots have always formed an essential part of such researches. Mobile robotics is a growing integrative field associated with several other fields like mechanical, electronic and electrical engineering, cognitive, social and computer sciences. It is associated with computational vision, robotics and artificial intelligence because of being involved in the design of automated systems. Mobile robotics is gradually becoming a stimulating area because of the efforts of various researchers. This book covers numerous topics by integrating contributions from many researchers around the globe. It lays stress on the computational techniques of programming mobile robots. It explores different aspects of theory and practice of robotics and consists of two sections: "Visual Perception, Mapping, Robot Localization, and Obstacle Avoidance" and "Path Planning and Motion Planning". It aims to serve as a guide to students and researchers studying mobile robots.

I would like to express my sincere thanks to the authors for their dedicated efforts in the completion of this book. I acknowledge the efforts of the publisher for providing constant support. Lastly, I would like to thank my family for their support in all academic endeavors.

Editor

Part 1

Visual Perception, Mapping, Robot Localization and Obstacle Avoidance

Mobile Robot Position Determination

Farouk Azizi and Nasser Houshangi
Purdue University Calumet
USA

1. Introduction

One of the most important reasons for the popularity of mobile robots in industrial manufacturing is their capability to move and operate freely. In order for the robots to perform to the expectations in manufacturing, their position and orientation must be determined accurately. In addition, there is a strong tendency to grant more autonomy to robots when they operate in hazardous or unknown environments which also requires accurate position determination. Mobile robots are usually divided into two categories of legged and wheeled robots. In this chapter, we focus on wheeled mobile robots.

Techniques used for position determination of wheeled mobile robots (or simply, mobile robots) are classified into two main groups: relative positioning and absolute positioning (Borenstein, 1996, 1997). In relative positioning, robot's position and orientation will be determined using relative sensors such as encoders attached to the wheels or navigation systems integrated with the robots. Absolute positioning techniques are referred to the methods utilizing a reference for position determination. The Global Positioning Systems, magnetic compass, active beacons are examples of absolute positioning systems.

Calculating position from wheel rotations using the encoders attached to the robot's wheels is called Odometry. Although odometry is the first and most fundamental approach for position determination, due to inherent errors, it is not an accurate method. As a solution to this problem, usually odometry errors are modeled using two different methods of benchmarks and multiple sensors. In this chapter, we will discuss odometry and two different methods to model and estimate odometry errors. At the end, an example for mobile robot position determination using multiple sensors will be presented.

2. Odometry

There is a consensus among researcher that odometry is the vital technique for mobile robot position determinations. The governing equations of odometry are based on converting rotational motion of robot wheels to a translational motion (Borenstein, 1996, 1997). Although odometry is an inexpensive method for position determination, it has several inherent issues. One issue is that errors accumulate over time and consequently make odometry unreliable over time. The odometry errors are mainly classified as systematic and nonsystematic (Borenstein, 1996). The source of systematic errors usually caused by:

- Average of both wheel diameters differs from nominal diameter
- Misalignment of wheels

- Uncertainty about the effective wheelbase (due to non-point wheel contact with the floor)
- Limited encoder resolution
- Limited encoder sampling rate

Nonsystematic errors are caused by the following conditions:

- Travel over uneven floors
- Travel over unexpected objects on the floor
- Wheel-slippage

Since these errors drastically affect the accuracy of odometry over short and long distances, it is empirical to accurately estimate the errors. The techniques used to overcome problems with odometry can be categorized to benchmark techniques and utilizing multiple sensors.

2.1 Correcting odometry using benchmark techniques

Benchmark techniques are based on testing robots over some predefined paths. In each experiment, the actual position and orientation of the robot is compared with the theoretical final position and orientation of the robot. Using the robot's kinematic equation a series of correction factors are derived to be incorporated in future position determination.

Borenstein and Feng proposed a test bench to model and estimate systematic errors (Borenstein, 1995). The method is called UMBmark and is based on running a robot with differential wheel drives on a square clockwise and counter clockwise paths for several times and compare the recording with the actual final position and orientation of the robot to generate the odometry errors. The errors are then classified as type A and type B. Type A error is the error that increases (or decreases) amount of rotation of the robot during the square path experiment in both clockwise and counter clockwise direction while type B error is the kind of error that increases (decreases) amount of rotation in one direction (clockwise or counter clockwise) and decreases (increases) amount of rotation in other direction (counter clockwise or clockwise). The robot's geometric relations are used to calculate the wheelbase and distance errors in terms of type A and B errors. Finally, the results are used to find two correction factors for right and left wheels of the robot to be applied to odometry calculation to improve the positioning system. The advantage of this method lies in the fact that mechanical inaccuracies such as wheel diameters can be modeled and compensated for. However, the method is only suitable to overcome systematic errors and does not estimate nonsystematic errors.

Chong and Kleeman modified UMBmark to model odometry positioning system more effectively by offering a mechanism to include odometry's nonsystematic errors (Chong, 1997). This approach uses robot's geometric and dynamical equation in a more fundamental way to generate an error covariance matrix for estimating the odometry errors. The technique also utilizes an unloaded wheel installed independently on linear bearings of each wheel to minimize wheel slippage and motion distortion. As a result of these modifications, odometry errors are minimized to a level that could be considered a zero-mean white noise signal. The other important characteristic of this method is that the errors calculated at each given time are uncorrelated to the next or from previous errors. The systematic errors are calibrated using UMBmark test. To model non-systematic errors, it is assumed that the robot motion can be divided to small segments of translational and rotational motions. For each segment the associated covariance matrix is calculated based on the physical model of the motion. The covariance matrix is updated using the previous calculation of the matrix. This method does not use any signal processing filter to estimate the errors. The main drawback of the method is the use of unloaded wheels along with drive wheels which limits the robot motion significantly.

Antonelli and coworkers proposed a method to calibrate odometry in mobile robots with differential drive (Antonelli, 2005). This method is based on estimating some of odometry parameters using linear identification problems for the parameters. These parameters come from kinematic equations of the odometry which are the sources of odometry's systematic errors. The estimation model is derived from the least square technique which allows a numerical analysis of the recorded data. Unlike UMBmark, this method is based on a more flexible platform and does not require a predefined path for the robot. Since the precision of the robot position and orientation at any given time is very important in estimating the robot position at the next moment, a vision-based measurement system is used to record robot's position and orientation. The experimental results shows this technique provide less error in estimating the robot position and orientation when compared to UMBMark technique.

Larsen and co-workers reported a method to estimate odometry systematic errors using Augmented Kalman Filter (Larsen, 1998). This technique estimates the wheel diameter and distance traveled with a sub-percent precision by augmenting the Kalman filter with a correction factor. The correction factor changes the noise covariance matrix to rely on the measurements more than the model by placing more confidence on the new readings. The augmented Kalman filter uses encoder readings as inputs and vision measurements as the observations. The only condition to achieve such precision is to use more sensitive measurement system to measure the robot position and orientation.

Martinelli and Siegwart proposed a method to estimate both systematic and nonsystematic errors odometry errors during the navigation (Martinelli, 2003). The systematic errors are estimated using augmented Kalman filter technique proposed in (Larsen, 1998) and the nonsystematic errors are estimated using another Kalman Filter. The second Kalman filter observations is called Observable Filter (OF) come from the estimates of the robot configuration parameters of the segmented Kalman filter. The main idea of Observable filters is to estimate the mean and variance of the observable variables of the robot parameters which are characterizing the odometry error.

2.2 Correcting odometry using multiple sensors

The other method to estimate the odometry errors is to integrate odometry with information from another sensor. The information from another sensor eventually is used to reset odometry errors especially during long runs. In many studies Global Positioning System (GPS), Inertial Navigation System (INS), compass, vision and sonar have been used in conjunction with odometry for position determination. In most cases Kalman filter or a derivation of Kalman filter, such as Indirect Kalman filter (IKF), Extended Kalman filter (EKF) and Unscented Kalman filter (UKF) has been used to integrate the information. In the following, we discuss few works have been done in this direction.

Park and coworkers employed a dead reckoning navigation system using differential encoders installed on the robot wheels and a gyroscope which is attached to robot (Park, 1997). The approach is based on estimation and compensation of the errors from the differential encoders and the gyroscope angular rate output. An Indirect Kalman filter (IKF) is used to integrate heading information from gyroscope and odometry. The output of IKF is used to compensate the errors associated with heading information in odometry as well as the error in gyroscope readings. The improved heading has been used in some formalism to give more accurate position and heading of the mobile robot. The work is followed by introducing a system of linear differential equations for each one of position errors on x- and y- directions, heading rate error, left and right wheel encoder errors, gyro scale factor error, and gyro bias drift error and wheel base error. The differential matrix derived from that linear differential equation is the equation that is used as the input for Indirect Kalman filter. The advantage of this method is that by including both the encoder and the gyroscope errors as the input of Kalman filter, both odometry and gyroscope errors can be estimated.

Cui and Ge in their work utilized Differential Global Positioning System (DGPS) as the basic positioning system for autonomous vehicle (Cui, 2001). GPS is currently used in many applications of land vehicle navigation and it is based on using information about the location and direction of motion of a moving vehicle which received from different satellites orbiting the earth. The expected accuracy of this technique is less than 0.5 meters. DGPS is similar to GPS but it uses two or more different satellite signals to localize a moving vehicle with more accuracy. One of the problems with both GPS and DGPS is when these methods are used to track a moving vehicle in an urban canyon. In such an environment, tall buildings prevent GPS to receive signals from the satellites. Therefore, it's critical to decrease the number of required satellite signals when the vehicle is moving in urban canyons. To achieve this goal, a constrained method was used that approximately modeled the path of the moving vehicle in an urban canyons environment as pieces of lines which is the cross section of two different surfaces. Each surface can be covered by two satellites at each moment. This will decrease the number of different satellite signals needed for vehicle localization to two. However, the method can switch back to use more than two satellite signals once the vehicle is not in moving in an urban canyon. In this method a pseudorange measurement is proposed to calculate distance from each satellite. Since the GPS receiver is modeled so that the measurement equation is nonlinear, the Extended Kalman Filter is used to the augmented matrix generated from the measurements to estimate the vehicle's position. While this work has not used odometry as the basic method for positioning of the vehicle, it decreases the number of required satellite signals to estimate the vehicle position when it travels in urban canyons. Recently Chae and coworkers have reported a method that uses EKF to efficiently integrate data from DGPS and INS sensors (Chae, 2010). The proposed method is designed to cover the task of mobile robot position determination during the times the robot navigates in urban areas with very tall buildings in which GPS signals are inaccessible.

Borenstein and Feng in another work have introduced a method called gyrodometry (Bronstein, 1996). In this method, the odometry is corrected for systematic errors using UMBmark. The approach intends also to get around non-systematic errors such as the errors generated when the robot traverses on a bump. The nonsystematic errors impact the robot for short period and is not detected by odometry.Once the robot faced a bump or another source of nonsystematic errors, odometry is replaced by gyro-based positioning system which doesn't fail in that situation and can provide a reliable information as the robot passes the bump. In practice both sets of information are available all the time and it is assumed that two sets of data, as long as there is no bump or object ahead of the robot, return almost the same information. In fact the procedure starts when odometry data differs substantially from gyro data. At the end of the time period robot traversing on a bump, the odometry is set by gyro positioning system. Afterwards, the odometry is the main positioning system and uses the correct new starting point. The gyro output drifts over time and this was compensated using the method proposed by Barshan and White (Barshan, 1993, 1994, 1995). Using this method both types of odometry errors are compensated and as result more reliable position and orientation are calculated. This approach is simple and doesn't need any sensor integration. While the method is reliable for the conditions it was tested for, it does not include the situation where odometry fails due to other nonsystematic errors such as moving on an uneven surface or in presence of wheel slippage.

Using ultrasonic beacons is another option to improve odometry which is proposed by Kleeman (Kleeman, 1992). In this approach active ultrasonic beacons are used to locate the robot at any given time. The information from beacons was used as a reliable position to reset the odometry. It is well known that ultrasonic measurements have random errors which don't accumulate over time but in contrast are not smooth and require to be compensated. In this method, an ultrasonic localizer is used which has six ultrasonic beacons and a transmitter controller which sequences the firing of the beacons in a cyclic manner. Beacons have 150 milliseconds intervals to have the previous pulse reverberation settled. Beacon 1 is distinguished by transmitting two bursts 3 milliseconds apart. The robot has an onboard receiver which is composed of eight ultrasonic receiver arranged at 45 degrees intervals. In modeling the positioning system, in addition to position of robot in both x- and y-directions, velocity of the robot, beacon cycle time, speed of sound and beacon firing time are introduced as the states of the system. An Iterative Extended Kalman Filter (IEKF) is utilized to integrate two sets of data from the beacons and from odometry. This method provides reliable information about the robot position in a constructed environment. However, it has several important drawbacks which make it unsuitable in many situations. For example, there is a problem with having delayed time arrival of an ultrasonic beacon due to an indirect path incorporating reflection off obstacles, walls, ceiling or floor. These echoed arrival time should be identified and must not be taken into account in estimation otherwise IEKF result in erroneous state estimation. In addition, one could easily point out that this method is appropriate only for indoor applications when ultrasonic beacons can be installed.

Barshan and Whyte have developed an Inertial Navigation system to navigate outdoor mobile robots (Barshan, 1993, 1994, 1995). In this system, three solid states gyroscope, a triaxial accelerometer and two Electrolevel tilt sensor are used. One of the important results of this work was the development of a method to model drift error of inertial sensors as an exponential function with time. Modeling the drift error for each sensor was done by leaving the sensor on a surface motionless. The sensor's readings were recorded until it was stabled. The sensor's readings are then modeled as an exponential function. The proposed Navigation system has been tested on radar-equipped land vehicles. The orientation from inertial sensors is reliable for 10 minutes. However, in the presence of electromagnetic fields information regarding the heading is valid only for 10 seconds. Although, using the inertial sensors reliably (after the drift errors were modeled) for 10 minutes is an improvement for vehicle position determination, it is not good enough when compared with systems which integrate two or more sensors. It is assumed that in all cases the errors associated with inertial sensor have been modeled and were taken in to account before the position and orientation of the vehicle was calculated.

Gyro's scale factor is provided by manufacturer to convert gyro output (digital or analog signal) to degrees per second. The scale factor is not a fixed and varies slightly with gyro's temperature, direction of rotation and rate of rotation. Therefore, modeling the variation of scale factor is critical to accurate calculation of gyro's angular information. Borenstein has used odometry and a gyroscope which was corrected for its scale factor variation along with its drift error (Borenstein, 1998). In addition, an Indirect Kalman Filter is used to model all errors corresponding with this method.

Roumeliotis and Bekey have utilized multiple sensors to navigate a mobile robot with two drive wheel and one cast (Roumeliotis, 2000). In this work one sensor (a potentiometer) was installed on rear wheel, to measure robot's steering angle. Additional sensors were: 1) two encoders on each wheel, 2) one single axis gyroscope, and 3) one sun sensor. A sun sensor is capable of measuring of the robot's absolute orientation based on sun position. Such sensor could be an effective alternative in applications such as Mars explorations where there is no access to GPS signals or strong magnetic field. It is indicated that sun sensor data should be used as often as one fifth of other sensors to achieve better estimation. Also this research shows that, by excluding robot's angular velocity and acceleration and translational acceleration into the estimation system state (in simulations), the system is very sensitive to changes in orientation caused by external sources such as slippage. Therefore to reduce this effect, robot angular velocity and angular acceleration as well as the robot translational acceleration are included into the estimation system's state.

Amarasinghe and co-workers have reported a method to integrate information from a laser range finder and a laser camera to navigate a mobile robot (Amarasinghe, 2010). In this technique, a camera is used to detect landmarks and the position of the landmark is measured by the laser range finder using laser camera. The information from two sensors is integrated in a Simultaneous Localization and Mapping (SLAM) platform supported by an Extended Kalman Filter (EKF). While the main advantage of this work is using appropriate sensors for detecting the landmarks and calculating the robot's position, it provides unreliable information in an indoor setting with no landmarks or in an unfavorable lighting condition such as a smoke-filled environment.

2.2.1 An example of using multiple sensors for position determination

In this section an example of position determination using sensor integration is discussed. The mobile considered is a wheeled robot with two driving wheels in the front and one dummy wheel in the back. The mobile robot position and orientation is represented by vector $\underline{P}(kT_s) = [p(kT_s), \psi(kT_s)]^T$ at time step kT_s, where k is a positive integer and T_s is the sampling period. For simplicity, T_s will be dropped from notation after this point and the mobile robot is assumed to be traveling on a flat surface. The vector $\underline{P}(k) = [p(k), \psi(k)]^T$ specifies the Cartesian coordinates of the robot, and the $\psi(k)$ defines the orientation (heading) of the robot. The position and orientation vector $\underline{P}(k)$ is updated in each sampling period during the robot's motion. Two coordinate frames are assigned as shown in figure 1. Body coordinate frame which is attached to the robot and moves with the robot. World coordinate frame, which sometimes is called navigation coordinate frame, is fixed at a reference point and the robot position is measured with respect to this frame. The superscript 'O' is used to denote the information obtained from odometry. The robot position $^{o}p(k)$ is determined incrementally by

$$^{o}\underline{p}(k+1) = {}^{o}\underline{p}(k) + R_{\psi o}(k)\underline{v}^{b}(k)T_s \qquad (1)$$

where $\underline{v}^b(k) = [v_x^b(k) \ v_y^b(k)]^T$, is the robot's velocity in the body coordinate frame, and $R_{\psi_o}(k)$ denotes the rotational transformation matrix from the body coordinate frame to the world coordinate frame defined by

$$R_{\psi_o}(k) = \begin{bmatrix} \cos\psi_o(k) & -\sin\psi_o(k) \\ \sin\psi_o(k) & \cos\psi_o(k) \end{bmatrix} \tag{2}$$

where $\psi_o(k)$ is heading of mobile robot based on odometry. The velocity component $v_y^b(k)$ is assumed to be zero because of forward motion of the mobile robot and $v_x^b(k)$ is calculated by

$$v_x^b(k) = \frac{1}{2}(v_{er}(k) + v_{el}(k)) \tag{3}$$

where $v_{er}(k)$ is the measured translational velocity of robot's right wheel and $v_{el}(k)$ is the left wheel's translational velocity. Figure 1 shows the geometric relations between robot's position and orientation in three consequent samples.

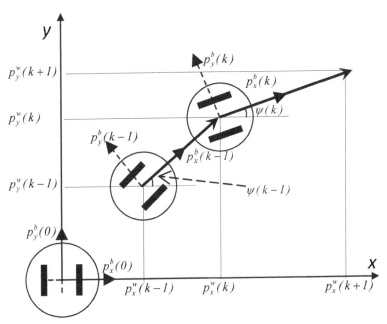

Fig. 1. Position determination based on dead reckoning.

The heading angle $\psi_o(k)$ is calculated by

$$\underline{v}^w(k) = R_{\psi_o}(k)\underline{v}^b(k) \tag{4}$$

$$\psi_o(k) = \tan^{-1}(v_y^w(k) / v_x^w(k)) \tag{5}$$

where $\underline{v}^w(k) = [v_x^w(k) \quad v_y^w(k)]^T$ is the translation velocity of mobile robot in world coordinate frame. The orientation based on odometry is updated by

$$\psi_o(k+1) = \psi_o(k) + \dot{\psi}_o(k)T_s \tag{6}$$

where the rate of change of orientation $\dot{\psi}_o(k)$ is calculated by

$$\dot{\psi}_o(k) = \frac{v_{er}(k) - v_{el}(k)}{b} \tag{7}$$

where b is the robot's wheelbase.

To alleviate the errors from odometry, the information from second sensor is integrated with odometry. Since based on the equations (1) and (4), the robot's heading plays an important role in position determination, a single axis gyroscope is used to determine the heading of the robot. The gyro's drift error is modeled using the technique introduced by Barshan (Barshan, 1993-1995). Figure 2 shows the gyro's output when the gyro examined on a flat and stationary surface for 14 hours. The result suggests that the gyro output increases exponentially when it was stationary during the test. Gyro's output consists of gyro accurate angle rate, the modeled error and a white noise:

$$\dot{\psi}_G(k) = \dot{\psi}_G^a(k) + e_m(k) + \eta_G(k) \tag{8}$$

where $\dot{\psi}_G^a(k)$ is actual heading change rate of mobile robot based on gyroscope reading, $\dot{\psi}_G(k)$ is heading change rate of mobile robot based on gyroscope reading, $e_m(k)$ is the gyroscope bias error, and $\eta_G(k)$ is the associate white noise.

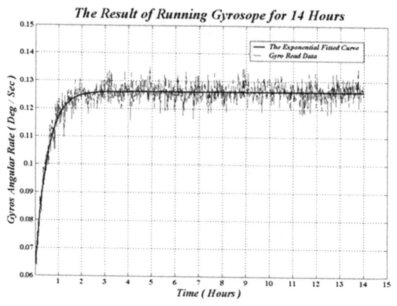

Fig. 2. The gyroscope's output for 14 hours.

A nonlinear parametric model for bias error was fitted to the data from gyroscope using least square fit method:

$$e_m(t) = C_1(1 - e^{-t/T}) + C_2 \tag{9}$$

$$\dot{e}_m(t) = \frac{C_1 + C_2}{T} + \frac{1}{T}e_m(t) \tag{10}$$

with initial conditions $e_m(0) = 0$ and $e_m(0) = C_1 / T$. After discretizing (10) becomes:

$$e_m(k+1) = \frac{T}{T + T_s}e_m(k) + \frac{T}{T + T_s}(C_1 + C_2) \tag{11}$$

The best fitting parameter value to experimental data obtained from gyroscope with zero input for 14 hours sampled every second are $C_1 = 0.06441$, $T = 30.65$ s, and $C_2 = 0.06332$.
If we assume that the original position is known, the next position of the robot in the world frame can be determined using equation (1) by replacing the robot's heading with the information calculated from the gyro.
The mobile robot position $^G p(k)$ in world coordinate frame based on gyro's readings is estimated by

$$^G\underline{p}^w(k+1) = \begin{bmatrix} ^G p_x^w(k+1) \\ ^G p_y^w(k+1) \end{bmatrix} = \begin{bmatrix} ^G p_x^w(k) \\ ^G p_y^w(k) \end{bmatrix} + \begin{bmatrix} \cos \psi_G^a(k) & -\sin \psi_G^a(k) \\ \sin \psi_G^a(k) & \cos \psi_G^a(k) \end{bmatrix} \begin{bmatrix} v_x^b(k) \\ v_y^b(k) \end{bmatrix} T_s \tag{12}$$

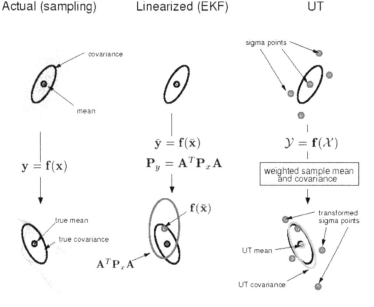

Fig. 3. Unscented Transformation (Wan, 2000).

The next step is to effectively integrate the information from odometry and gyroscope so that the odometry errors can be accurately estimated. As shown, the Kalman filter gives optimal state estimation to a linear system (Kalman, 1982). The EKF is an extension of the Kalman filter for nonlinear functions. The EKF is based on linearizing the nonlinear functions and applying Kalman filter for the estimation. Due to nonlinear nature of errors associated with odometry and other sensors, many studies have used Extended Kalman filter to integrate two or more sets of data for more accurate position determination. However, since the level of nonlinearity for odometry's nonsystematic error is high, the EKF may not be the best choice. Unscented Kalman filter (UKF) is another extension for Kalman filter which can provide accurate estimation for systems with high nonlinearity such as our system.

An important aspect of the UKF is that it can be applied to nonlinear functions without a need for linearizing (Julier, 1995, 1996). It is much simpler than EKF to implement because it doesn't require the calculation of Jacobian matrix which may provide some difficulties in the implementation.

The UKF is based on Unscented Transform (UT) that transforms the domain of the nonlinear function to another space of sigma points that has the same statistics as the original points (Julier, 2000). The UT is capable of estimating the statistics of a random variable defined by a nonlinear function (Julier, 2000). Figure 3 shows the general idea of the UT (Wan, 2000). In this transformation, a set of points called sigma point, are chosen such that it has mean of $\overline{x}$ and covariance of P_x. The first two moments of x can be calculated by choosing $2n+1$ sample points (also called sigma points) as follow

$$\chi_0(k|k) = \overline{x} \tag{13}$$

$$\chi_i(k|k) = \overline{x} + (\sqrt{(n+\kappa)P_x})_i \quad i = 1,..,n \tag{14}$$

$$\chi_i(k|k) = \overline{x} - (\sqrt{(n+\kappa)P_x})_i \quad i = n+1,..,2n \tag{15}$$

$$W_0 = \kappa/(n+\kappa) \tag{16}$$

$$W_i = 1/\{2(n+\kappa)\} \quad i = 1,..,2n \tag{17}$$

where κ is scaling factor, $(\sqrt{(n+\kappa)P_x})_i$ is the i-th row or column of matrix square root of $(n+\kappa)P_x$ and W_i s are the weight associated with i-th sigma point. It is known that

$$\sum_{i=0}^{2n} W_i = 1 \tag{18}$$

The sigma point are propagated through the nonlinear function of $f(x)$ as

$$Y_i = f(\chi_i(k|k)) \quad i = 0,..,2n \tag{19}$$

Mean of output of function $f(x)$ can be determined as

$$\overline{y} = f(Y_i) \tag{20}$$

Therefore, it is possible to estimate the statistics of a nonlinear function without taking derivatives from the function as it is needed in the case of EKF. This makes implementation much easier than EKF which needs linearizing of the nonlinear function around the operating points and calculation of the Jacobian matrices. The parameter κ provides another degree of freedom to eliminate the estimation error for higher order statistics. It can be positive or negative but choosing a negative number returns a non-positive semi-definite estimation for P_{yy} (Wan, 2000).

Fig. 4. The mobile robot with gyroscope.

Julier and Uhlmann (Julier, 1995, 1996) were first to introduce the UT as a technique for the estimation of nonlinear systems. Their work was based on the intuition that, with a fixed number of parameters it should be easier to approximate a Gaussian distribution than it is to approximate an arbitrary nonlinear function or transformation. This could be easily drawn from the way the sigma points of the UT are being calculated. There are few versions for the UKF which they differ from each other based on whether to include the random noise into the calculations.

Houshangi and Azizi have used the UKF to successfully integrate information from odometry and a single axis gyroscope (Houshangi, 2005, 2006; Azizi, 2004). The mobile robot used in their work is a Pioneer 2-Dxe from ActivMedia Robotics Corporation. Figure 4 illustrates the robot and the single axis gyroscope used as the second sensor for position determination. The robot is a two-wheel drive and incremental encoders are installed on each wheel. These wheels have pneumatic rubber tires which give better mobility but potentially are additional source of error for positioning based on odometry such as inequality of wheel diameters which inescapable. One rear caster is used to stabilize the robot's motion and standing.

The difference between robot position and orientation comes from odometry and robot position and orientation are calculated by

$$e_x(k) = {}^G P_x^w(k) - {}^O P_x^w(k) \tag{21-1}$$

$$e_y(k) = {}^G P_y^w(k) - {}^O P_y^w(k) \tag{21-2}$$

$$e_\psi(k) = \psi_G^a(k) - \psi_o(k) \tag{21-3}$$

Each one of these errors has a modeled part and an associate noise as shown in equations (20):

$$e_x(k) = e_{mx}(k) + \eta_x(k) \tag{22-1}$$

$$e_y(k) = e_{my}(k) + \eta_y(k) \tag{22-2}$$

$$e_\psi(k) = e_{m\psi}(k) + \eta_\psi(k) \tag{22-3}$$

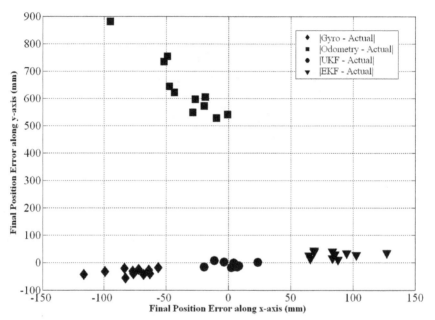

Fig. 5. Robot's position along the x-axis.

Three independent UKF described in equations (11-18) are applied to each one of the errors defined in equation (22). The estimated errors are included when an accurate position and orientation for the robot were calculated in the next sampling interval. The results were compared to the results of applying EKF to equation (22).

To evaluate the approach an experiment was designed so that the robot was run over a 3-meter square for eleven runs. The final position of the robot was calculated with respect to the actual final position. As shown in Figure 5, the UKF returns more accurate results compared to the EKF and better than using odometry or gyroscope alone. This is partly due the nature of odometry errors which are nonlinear and can be estimated more effectively using UKF rather than EKF.

3. Conclusions

Mobile robot position determination is vital for the effectiveness of robot's operation regardless of the task. Odometry is the fundamental technique for position determination but with many issues described in this chapter. The odometry's errors can be estimated using benchmark techniques and utilizing multiple sensors. Kalman Filtering is the most popular technique to integrate information from multiple sensors. However, most of physical and engineering phenomena are nonlinear and it is necessary to modify Kalman filter to estimate nonlinear systems. The EKF is able to deal with nonlinear systems and is used for mobile robot position determination. As was discussed, the odometry errors have high nonlinearity in nature. The newer extension of Kalman Filter using a transform called Unscented Transform provides another alternative for sensor integration. An example was provided using UKF to integrate gyro data and odometry. The implementation results indicated utilizing multiple sensors using UKF provided more accurate position even as compared to EKF.

4. References

Borenstein, J. & Feng. L. (1996) Measurement and Correction of Systematic Odometry Errors in Mobile Robots. *IEEE Journal of Robotics and Automation*, pp. 869-880, Vol 12, No 6, (December 1996).

Borenstein, J., Everett, H.R., Feng, L., & Wehe, D. (1997) Mobile Robot Positioning: Sensors and Techniques. *Invited paper for the* Journal of Robotic Systems, Special Issue on Mobile Robots. pp. 231-249, Vol. 14, No. 4, (April 1997).

Borenstein, J. & Feng. L. (1995) UMBmark: A Benchmark Test for Measuring Odometry Errors in Mobile Robots. *Proceedings of SPIE Conference on Mobile Robots*, Philadelphia, (October, 1995).

Chong, K.S. & Kleeman, L. (1997) Accurate Odometry and Error Modeling for a Mobile Robot. *Proceeding of the, IEEE Intl. Conference on Robotics and Automation*, Albuquerque, New Mexico, pp. 2783-2788, (April 1997)

Antonelli, G, Chiaverini, S. & Fusco, G. (2005) A calibration method for odometry of mobile robots based on the least-squares technique: theory and experimental validation. *IEEE Transactions on Robotics*, vol. 21, pp. 994–1004, (Oct. 2005).

Larsen, T. D., Bak, M., Andersen, N. A. & Ravn, O. (1998) Location estimation for an autonomously guided vehicle using an augmented Kalman filter to autocalibrate the odometry. *First International Conference on Multisource-Multisensor Information Fusion (FUSION'98)*, (1998).

Martinelli, R. & Siegwart, R. (2003) Estimating the Odometry Error of a Mobile Robot During Navigation *European Conference on Mobile Robots*, Poland (2003).

Park, K. C., Chung, H., Choi, J., and Lee, J.G. (1997) Dead Reckoning Navigation for an Autonomous Mobile Robot Using a Differential Encoder and Gyroscope", Proc. of the 8th Intl. Conference on Advanced Robotics, Monterey, CA, July 1997, 441-446

Cui, Y.J. & Ge, S.S. (2001) Autonomous vehicle positioning with GPS in urban canyon environments. *Proc. of IEEE International Conference on Robotics and Automation*, pp. 1105-1110, Seoul, Korea (Feb. 2001)

Chae, H., Christiand, Choi, S., Yu W & Cho, J. (2010) Autonomous Navigation of Mobile Robot based on DGPS/INS Sensor Fusion by EKF in Semi-outdoor Structured Environment. *Proceedings of the IEEE/RSJ International Conference on Intelligent Robots and Systems (IROS)*, pp. 1222-1227 (Oct. 2010)

Bronstein, J. & Feng, L. (1996) Gyrodometry: A New Method for Combining Data from Gyros and Odometry in Mobile robots. *Proceeding of the 1996 IEEE International*

Conference on Robotics and Automation, pp. 423–428, Minneapolis, Minnesota, (April 1996)

Kleeman, L. (1992) Optimal estimation of position and heading for mobile robots using ultrasonic beacons and dead-reckoning. *IEEE International Conference on Robotics and Automation*, Nice, France, pp 2582-2587, (May 1992)

Barshan, B., and Durrant–Whyte, H. F. (1993) An Inertial Navigation System for a for Mobile Robot. *Proceeding of 1993 IEEE/RSJ International Conference on Intelligent Robots and Systems*, pp. 2243 – 2247, Yokohama, Japan, (July 1993)

Barshan, B., and Durrant–Whyte, H. F. (1994) Evaluation of a Solid – State Gyroscope for Mobile Robots Application *IEEE Transactions on Instrumentation and Measurement*, pp. 61 – 67, Vol. 44, No. 1, (February 1994)

Barshan, B. & Durrant-Whyte, H. F. (1995)Inertial Navigation Systems for Mobile Robots. *IEEE Transactions on Robotics and Automation*, pp. 328 – 342, Vol. 11, No. 3, (June 1995)

Borenstein, J. (1998) Experimental Evaluation of the Fiber Optics Gyroscope for Improving Dead-reckoning Accuracy in Mobile Robots *Proceeding Of the IEEE International Conference on Robotics and Automation*, 3456 – 3461, Leuven, Belgium, (May 1998)

Roumeliotis, S.I. & Bekey, G.A. (2000) Bayesian Estimation and Kalman Filtering: A unified Framework for Mobile Robot Localization. *Proceedings of IEEE International Conference on Robotics and Automation*, pp. 2985-2992, San Fransisco, (April 2000).

Amarasinghe, D., Mann, G.K.I. & Gosine, R.G. (2010) Landmark detection and localization for mobile robot applications: a multisensor approach. *Robotica*, pp. 663–673 (2010) volume 28.

Kalman, R. E. (1982) A New Approach to Linear Filtering and Prediction Problems. *Transaction of the ASME – Journal of Basic Engineering, (Series D)*, pp. 35 – 45.

Julier, S. J. & Uhlmann, J. K. &Durrant (1995) Whyte A New Approach for Filtering Nonlinear Systems. *Proceeding of the American Control Conference*, pp. 1628 – 1632, Seattle, Washington, 1995.

Julier, S. J. & Uhlmann, J. K. (1996) A New Method for Approximating Nonlinear Transformation of Probability Distributions. *Tech. Report, Robotic Research Group, Department of Engineering Science*, University of Oxford, 1996

Julier, S. J. & Uhlmann, J. K. (1997) A New Extension of the Kalman Filter to Nonlinear Systems. Proceeding of Aerospace: *The 11th International Symposium on Aerospace/ Defense Sensing, Simulation and Controls*, pp. 182 – 193, Orlando, Florida, (2000)

Julier, S. J. (2002) The scaled unscented transformation. *Proceedings of the American Control Conference*, Vol. 6, pp. 4555-4559, (2002)

Wan, E.A. & van der Merwe, R. (2000) The Unscented Kalman Filter for Nonlinear Estimation. *Proceedings of Symposium on Adaptive Systems for Signal Processing, Communication and Control*, Lake Louise, Alberta, Canada, (Oct. 2000)

Azizi, F. & Houshangi, N. (2004) Mobile robot position determination using data from gyro and odometry. *Canadian Conference on Electrical Engineering*, Vol. 2, pp.719 – 722, (May 2004).

Houshangi, N. & Azizi, F. (2005) Accurate mobile robot position determination using unscented Kalman filter. *Canadian Conference on Electrical Engineering*, pp. 846 – 851, (May 2005)

Houshangi, N. & Azizi, F. (2006) Mobile Robot Position Determination Using Data Integration of Odometry and Gyroscope. *World Automation Congress*. (2006), pp. 1–8

2

Development of an Autonomous Visual Perception System for Robots Using Object-Based Visual Attention

Yuanlong Yu, George K. I. Mann and Raymond G. Gosine
Faculty of Engineering and Applied Science, Memorial University of Newfoundland
St. John's, NL,
Canada

1. Introduction

Unlike the traditional robotic systems in which the perceptual behaviors are manually designed by programmers for a given task and environment, autonomous perception of the world is one of the challenging issues in the cognitive robotics. It is known that the *selective attention* mechanism serves to link the processes of perception, action and learning (Grossberg, 2007; Tipper et al., 1998). It endows humans with the *cognitive capability* that allows them to *learn* and *think* about how to perceive the environment autonomously. This visual attention based autonomous perception mechanism involves two aspects: *conscious aspect* that directs perception based on the current task and learned knowledge, and *unconscious aspect* that directs perception in the case of facing an unexpected or unusual situation. The *top-down attention* mechanism (Wolfe, 1994) is responsible for the conscious aspect whereas the *bottom-up attention* mechanism (Treisman & Gelade, 1980) corresponds to the unconscious aspect. This paper therefore discusses about how to build an artificial system of autonomous visual perception.

Three fundamental problems are addressed in this paper. The first problem is about pre-attentive segmentation for object-based attention. It is known that attentional selection is either *space-based* or *object-based* (Scholl, 2001). The space-based theory holds that attention is allocated to a spatial location (Posner et al., 1980). The object-based theory, however, posits that some pre-attentive processes serve to segment the field into discrete objects, followed by the attention that deals with one object at a time (Duncan, 1984). This paper proposes that object-based attention has the following three advantages in terms of computations: 1) Object-based attention is more robust than space-based attention since the attentional activation at the object level is estimated by accumulating contributions of all components within that object, 2) attending to an exact object can provide more useful information (e.g., shape and size) to produce the appropriate actions than attending to a spatial location, and 3) the discrete objects obtained by pre-attentive segmentation are required in the case that a global feature (e.g., shape) is selected to guide the top-down attention. Thus this paper adopts the object-based visual attention theory (Duncan, 1984; Scholl, 2001).

Although a few object-based visual attention models have been proposed, such as (Sun, 2008; Sun & Fisher, 2003), developing a pre-attentive segmentation algorithm is still a challenging issue as it is a unsupervised process. This issue includes three types of challenges: 1) The ability to automatically determine the number of segments (termed as *self-determination*), 2) the computational efficiency, and 3) the robustness to noise. Although K-labeling methods (e.g., normalized cut (Shi & Malik, 2000)) can provide the accuracy and robustness, they are ineffective and inefficient when the number of segments is unknown. In contrast, recent split-and-merge methods (e.g., irregular pyramid based segmentation (Sharon et al., 2006)) are capable of determining the number of segments and computationally efficient, whereas they are not robust to noise. This paper proposes a new pre-attentive segmentation algorithm based on the irregular pyramid technique in order to achieve the self-determination and robustness as well as keep the balance between the accuracy and efficiency.

The second problem is about how to model the attentional selection, i.e., model the cognitive capability of *thinking* about what should be perceived. Compared with the well-developed bottom-up attention models (Itti & Baldi, 2009; Itti et al., 1998), modeling the top-down attention is far from being well-studied. The top-down attention consists of two components: 1) Deduction of task-relevant object given the task and 2) top-down biasing that guides the focus of attention (FOA) to the task-relevant object. Although some top-down methods have been proposed, such as (Navalpakkam & Itti, 2005), several challenging issues require further concerns. Since the first component is greatly dependent on the knowledge representation, it will be discussed in the next paragraph. Regarding the second component, the first issue is about the effectiveness of top-down biasing. The main factor that decays the effectiveness is that the task-relevant object shares some features with the distracters. It indicates that the top-down biasing method should include a mechanism to make sure that the task-relevant object can be discriminated from distracters. The second issue is about the computational efficiency based on the fact that the attention is a fast process to select an object of interest from the image input. Thus it is reasonable to use some low-level features rather than high-level features (e.g., the iconic representation (Rao & Ballard, 1995a)) for top-down biasing. The third one is the adaptivity to automatically determine which feature(s) is used for top-down biasing such that the requirement of manually re-selecting the features for different tasks and environment is eliminated. This paper attempts to address the above issues by using the integrated competition (IC) hypothesis (Duncan et al., 1997) since it not only summarizes a theory of the top-down attention, which can lead to a computational model with effectiveness, efficiency and adaptivity, but also integrates the object-based attention theory. Furthermore, it is known that bottom-up attention and top-down attention work together to decide the attentional selection, but how to combine them is another challenging issue due to the multi-modality of bottom-up saliency and top-down biases. A promising approach to this issue is setting up a unified scale at which they can be combined.

The third problem is about the cognitive capability of autonomously *learning* the knowledge that is used to guide the conscious perceptual behavior. According to the psychological concept, the memory used to store this type of knowledge is called long-term memory (LTM). Regarding this problem, the following four issues are addressed in this paper. The first issue is about the unit of knowledge representations. Object-based vision theory (Duncan, 1984; Scholl, 2001) indicates that a general way of organizing the visual scene is to parcel it into discrete objects, on which perception, action and learning perform. In other words, the internal attentional representations are in the form of objects. Therefore objects are used as the units of the learned knowledge. The second issue is what types of knowledge should be modeled for guiding the conscious perceptual behavior. According to the requirements of the attention mechanism, this paper proposes that the knowledge mainly includes LTM task representations and LTM object representations. The *LTM task representation* embodies the association between the attended object at the last time and predicted task-relevant object at the current time. In other words, it tells the robot what should be perceived at each time. Thus its objective is to deduce the task-relevant object given the task in the attentional selection stage. The *LTM object representation* embodies the properties of an object. It has two objectives: 1) Directing the top-down biasing given the task-relevant object and 2) directing the post-attentive perception and action selection. The third issue is about how to build their structure in order to realize the objectives of these two representations. This paper employs the *connectionist approach* to model both representations as the self-organization can be more effectively achieved by using the cluster-based structure, although some symbolic approaches (Navalpakkam & Itti, 2005) have been proposed for task representations. The last issue is about how to learn both representations through the duration from an infant robot to a mature one. It indicates that a dynamic, constructive learning algorithm is required to achieve the self-organization, such as generation of new patterns and re-organization of existing patterns. Since this paper focuses on the perception process, only the learning of LTM object representations is presented.

The remainder of this paper is organized as follows. Some related work of modeling visual attention and its applications in robotic perception are reviewed in section 2. The framework of the proposed autonomous visual perception system is given in section 3. Three stages of this proposed system are presented in section 4, section 5 and section 6 respectively. Experimental results are finally given in section 7.

2. Related work

There are mainly four psychological theories of visual attention, which are the basis of computational modeling. Feature integration theory (FIT) (Treisman & Gelade, 1980) is widely used for explaining the space-based bottom-up attention. The FIT asserts that the visual scene is initially coded along a variety of feature dimensions, then attention competition performs in a location-based serial fashion by combining all features spatially, and focal attention finally provides a way to integrate the initially separated features into a whole object. Guided search model (GSM) (Wolfe, 1994) was further proposed to model the space-based top-down attention mechanism in conjunction with bottom-up attention. The GSM posits that the top-down request for a given feature will activate the locations that might contain the given feature. Unlike FIT and GSM, the biased competition (BC) hypothesis (Desimone & Duncan, 1995) asserts that attentional selection, regardless of being space-based or object-based, is a biased competition process. Competition is biased in part by the bottom-up mechanism that favors a local inhomogeneity in the spatial and temporal context and in part by the top-down mechanism that favors items relative to the current task. By extending the BC hypothesis, the IC hypothesis (Duncan, 1998; Duncan et al., 1997) was further presented to explain the object-based attention mechanism. The IC hypothesis holds that any property of an object can be used as a task-relevant feature to guide the top-down attention and the whole object can be attended once the task-relevant feature successfully captures the attention.

A variety of computational models of space-based attention for computer vision have been proposed. A space-based bottom-up attention model was first built in (Itti et al., 1998). The surprise mechanism (Itti & Baldi, 2009; Maier & Steinbach, 2010) was further proposed to model the bottom-up attention in terms of both spatial and temporal context. Itti's model was further extended in (Navalpakkam & Itti, 2005) by modeling the top-down attention mechanism. One contribution of Navalpakkam's model is the symbolic knowledge representations that are used to deduce the task-relevant entities for top-down attention. The other contribution is the multi-scale object representations that are used to bias attentional selection. However, this top-down biasing method might be ineffective in the case that environment contains distracters which share one or some features with the target. Another model that selectively tunes the visual processing networks by a top-down hierarchy of winner-take-all processes was also proposed in (Tsotsos et al., 1995). Some template matching methods such as (Rao et al., 2002), and neural networks based methods, such as (Baluja & Pomerleau, 1997; Hoya, 2004), were also presented for modeling top-down biasing. Recently an interesting computational method that models attention as a Bayesian inference process was reported in (Chikkerur et al., 2010). Some space-based attention model for robots was further proposed in (Belardinelli & Pirri, 2006; Belardinelli et al., 2006; Frintrop, 2005) by integrating both bottom-up and top-down attention.

Above computational models direct attention to a spatial location rather than a perceptual object. An alternative, which draws attention to an object, has been proposed by (Sun & Fisher, 2003). It presents a computational method for grouping-based saliency and a hierarchical framework for attentional selection at different perceptual levels (e.g. a point, a region or an object). Since the pre-attentive segmentation is manually achieved in the original work, Sun's model was further improved in (Sun, 2008) by integrating an automatic segmentation algorithm. Some object-based visual attention models (Aziz et al., 2006; Orabona et al., 2005) have also been presented. However, the top-down attention is not fully achieved in these existing object-based models, e.g., how to get the task-relevant feature is not realized.

Visual attention has been applied in several robotic tasks, such as object recognition (Walther et al., 2004), object tracking (Frintrop & Kessel, 2009), simultaneous localization and mapping (SLAM) (Frintrop & Jensfelt, 2008) and exploration of unknown environment (Carbone et al., 2008). A few general visual perception models (Backer et al., 2001; Breazeal et al., 2001) are also presented by using visual attention. Furthermore, some research (Grossberg, 2005; 2007) has proposed that the adaptive resonance theory (ART) (Carpenter & Grossberg, 2003) can predict the functional link between attention and processes of consciousness, learning, expectation, resonance and synchrony.

3. Framework of the proposed system

The proposed autonomous visual perception system involves three successive stages: pre-attentive processing, attentional selection and post-attentive perception. Fig. 1 illustrates the framework of this proposed system.

Stage 1: The *pre-attentive processing stage* includes two successive steps. The first step is the extraction of pre-attentive features at multiple scales (e.g., nine scales for a 640×480 image). The second step is the pre-attentive segmentation that divides the scene into proto-objects in an unsupervised manner. The *proto-objects* can be defined as uniform regions such that the pixels in the same region are similar. The obtained proto-objects are the fundamental units of attentional selection.

Stage 2: The *attentional selection stage* involves four modules: bottom-up attention, top-down attention, a combination of bottom-up saliency and top-down biases, as well as estimation of proto-object based attentional activation. The bottom-up attention module aims to model the unconscious aspect of the autonomous perception. This module generates a probabilistic location-based bottom-up saliency map. This map shows the conspicuousness of a location compared with others in terms of pre-attentive features. The top-down attention module aims

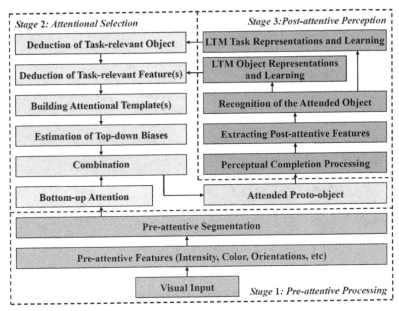

Fig. 1. The framework of the proposed autonomous visual perception system for robots.

to model the conscious aspect of the autonomous perception. This module is modeled based on the IC hypothesis and consists of four steps. *Step 1* is the deduction of the task-relevant object from the corresponding LTM task representation given the task. *Step 2* is the deduction of the task-relevant feature dimension(s) from the corresponding LTM object representation given the task-relevant object. *Step 3* is to build the attentional template(s) in working memory (WM) by recalling the task-relevant feature(s) from LTM. *Step 4* is to estimate a probabilistic location-based top-down bias map by comparing attentional template(s) with corresponding pre-attentive feature(s). The obtained top-down biases and bottom-up saliency are combined in a probabilistic manner to yield a location-based attentional activation map. By combining location-based attentional activation within each proto-object, a proto-object based attentional activation map is finally achieved, based on which the most active proto-object is selected for attention.

Stage 3: The main objective of the *post-attentive perception stage* is to interpret the attended object in more detail. The detailed interpretation aims to produce the appropriate action and learn the corresponding LTM object representation at the current time as well as to guide the top-down attention at the next time. This paper introduces four modules in this stage. The first module is perceptual completion processing. Since an object is always composed of several parts, this module is required to perceive the complete region of the attended object post-attentively. In the following text, the term *attended object* is used to represent one or all of the proto-objects in the complete region being attended. The second module is the extraction of post-attentive features that are a type of representation of the attended object in WM and used for the following two modules. The third module is object recognition. It functions as a decision unit that determines to which LTM object representation and/or to which instance of that representation the attended object belongs. The fourth module is learning of LTM object representations. This module aims to develop the corresponding LTM representation of the attended object. The probabilistic neural network (PNN) is used to build the LTM object representation. Meanwhile, a constructive learning algorithm is also proposed. Note that the LTM task representation is another important module in the post-attentive perception stage. Its learning requires the perception-action training pairs, but this paper focuses on the perception process rather than the action selection process. So this module will be discussed in the future work.

4. Pre-attentive processing

4.1 Extraction of pre-attentive features

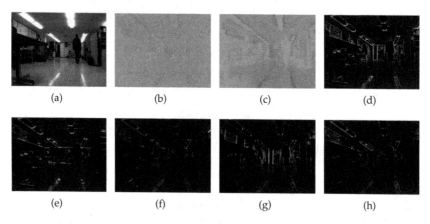

Fig. 2. Pre-attentive features at the original scale. (a) Intensity. (b) Red-green. (c) Blue-yellow. (d) Contour. (e) - (h) Orientation energy in direction $0°$, $45°$, $90°$ and $135°$ respectively. Brightness represents the energy value.

By using the method in Itti's model (Itti et al., 1998), pre-attentive features are extracted at multiple scales in the following dimensions: intensity $\mathbf{F}_{int}$, red-green $\mathbf{F}_{rg}$, blue-yellow $\mathbf{F}_{by}$, orientation energy $\mathbf{F}_{0_\theta}$ with $\theta \in \{0°, 45°, 90°, 135°\}$, and contour $\mathbf{F}_{ct}$. Symbol $\mathbf{F}$ is used to denote pre-attentive features.

Given 8-bit RGB color components $\mathbf{r}$, $\mathbf{g}$ and $\mathbf{b}$ of the input image, intensity and color pairs at the original scale are extracted as : $\mathbf{F}_{int} = (\mathbf{r} + \mathbf{g} + \mathbf{b})/3$, $\mathbf{F}_{rg} = \mathbf{R} - \mathbf{G}$, $\mathbf{F}_{by} = \mathbf{B} - \mathbf{Y}$, where $\mathbf{R} = \mathbf{r} - (\mathbf{g} + \mathbf{b})/2$, $\mathbf{G} = \mathbf{g} - (\mathbf{r} + \mathbf{b})/2$, $\mathbf{B} = \mathbf{b} - (\mathbf{r} + \mathbf{g})/2$, and $\mathbf{Y} = (\mathbf{r} + \mathbf{g})/2 - |\mathbf{r} - \mathbf{g}|/2 - \mathbf{b}$. Gaussian pyramid (Burt & Adelson, 1983) is used to create the multi-scale intensity and color pairs. The multi-scale orientation energy is extracted using the Gabor pyramid (Greenspan et al., 1994). The contour feature $\mathbf{F}_{ct}(s)$ is approximately estimated by applying a pixel-wise maximum operator over four orientations of orientation energy: $\mathbf{F}_{ct}(s) = \max_{\theta \in \{0°, 45°, 90°, 135°\}} \mathbf{F}_{0_\theta}(s)$, where s denotes the spatial scale. Examples of the extracted pre-attentive features have been shown in Fig. 2.

4.2 Pre-attentive segmentation

This paper proposes a pre-attentive segmentation algorithm by extending the irregular pyramid techniques (Montanvert et al., 1991; Sharon et al., 2000; 2006). As shown in Fig. 3, the pre-attentive segmentation is modeled as a hierarchical accumulation procedure, in which each level of the irregular pyramid is built by accumulating similar local nodes at the level below. The final proto-objects emerge during this hierarchical accumulation process as they are represented by single nodes at some levels. This accumulation process consists of four procedures.

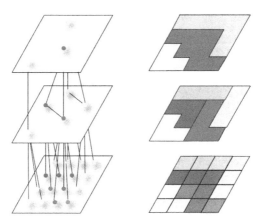

Fig. 3. An illustration of the hierarchical accumulation process in the pre-attentive segmentation. This process is shown from bottom to top. In the left figure, this process is represented by vertices and each circle represents a vertex. In the right figure, this process is represented by image pixels and each block represents an image pixel. The color of each vertex and block represents the feature value. It can be seen that the image is partitioned into three irregular regions once the accumulation process is finished.

Procedure 1 is decimation. A set of surviving nodes (i.e., parent nodes) is selected from the son level to build the parent level. This procedure is constrained by the following two rules (Meer, 1989): 1) Any two neighbor son nodes cannot both survive to the parent level and 2) any son node must have at least one parent node. Instead of the *random values* used in the stochastic pyramid decimation (SPD) algorithm (Jolion, 2003; Meer, 1989), this paper proposes a new recursive similarity-driven algorithm (i.e., the first extension), in which a son node will survive if it has the maximum *similarity* among its neighbors with the constraints of the aforementioned rules. The advantage is the improved segmentation performance since the nodes that can greatly represent their neighbors deterministically survive. As the second extension, *Bhattacharyya distance* (Bhattacharyya, 1943) is used to estimate the similarity between nodes at the same level (i.e., the strength of intra-level edges). One advantage is that the similarity measure is approximately scale-invariant during the accumulation process since Bhattacharyya distance takes into account the correlations of the data. The other advantage is that the probabilistic measure can improve the robustness to noise.

In *procedure 2*, the strength of inter-level edges is estimated. Each son node and its parent nodes are linked by inter-level edges. The strength of these edges is estimated in proportion to the corresponding intra-level strength at the son level by using the method in (Sharon et al., 2000).

Procedure 3 aims to estimate the aggregate features and covariances of each parent node based on the strength of inter-level edges by using the method in (Sharon et al., 2000).

The purpose of *procedure 4* is to search for neighbors of each parent node and simultaneously estimate the strength of intra-level edges at the parent level. As the third extension, a new

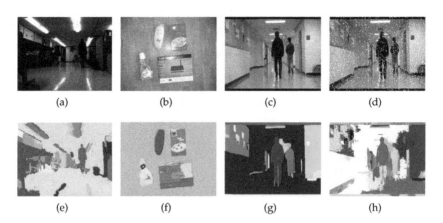

Fig. 4. Results of pre-attentive segmentation. (a)-(d) Original images, where (d) includes salt and pepper noise (noise density:0.1, patch size: 5×5 pixels). (e)-(h) Segmentation results. Each color represents one proto-object in these results.

neighbor search method is proposed by considering not only the graphic constraints but also the similarity constraints. A candidate node is selected as a neighbor of a center node if the similarity between them is above a predefined threshold. Since the similarity measure is scale-invariant, a fixed value of the threshold can be used for most pyramidal levels. The advantage of this method is the improved segmentation performance since the connections between nodes that are located at places with great transition are deterministically cut.

In the case that no neighbors are found for a node, it is labeled as a new proto-object. The construction of the full pyramid is finished once all nodes at a level have no neighbors. The membership of each node at the base level to each proto-object is iteratively calculated from the top pyramidal level to the base level. According to the membership, each node at the base level is finally labeled. The results of the pre-attentive segmentation are shown in Fig. 4.

5. Attentional selection

5.1 Bottom-up attention

The proposed bottom-up attention module is developed by extending Itti's model (Itti et al., 1998). Center-surround differences in terms of pre-attentive features are first calculated to simulate the competition in the spatial context:

$$\mathbf{F}'_f(s_c, s_s) = |\mathbf{F}_f(s_c) \ominus \mathbf{F}_f(s_s)| \tag{1}$$

where $\ominus$ denotes across-scale subtraction, consisting of interpolation of each feature at the surround scale to the center scale and point-by-point difference, $s_c = \{2, 3, 4\}$ and $s_s = s_c + \delta$ with $\delta = \{3, 4\}$ represent the center scales and surround scales respectively, $f \in \{int, rg, by, o_\theta, ct\}$ with $\theta \in \{0°, 45°, 90°, 135°\}$, and $F'_f(s_c, s_s)$ denotes a center-surround difference map.

These center-surround differences in terms of the same feature dimension are then normalized and combined at scale 2, termed as *working scale* and denoted as s_{wk}, using across-scale addition to yield a location-based conspicuity map of that feature dimension:

$$\mathbf{F}_f^s = \mathcal{N}\left(\frac{1}{6} \bigoplus_{s_c=2}^{4} \bigoplus_{s_s=s_c+3}^{s_c+4} \mathcal{N}\left(\mathbf{F}_f'(s_c, s_s)\right)\right) \tag{2}$$

where $\mathcal{N}$ is the normalization operator, $\bigoplus$ is across-scale addition, consisting of interpolation of each normalized center-surround difference to the working scale and point-by-point addition, $f \in \{int, rg, by, o_\theta, ct\}$, and $\mathbf{F}_f^s$ denotes a location-based conspicuity map.

All conspicuity maps are point-by-point added together to yield a location-based bottom-up saliency map $\mathbf{S}_{bu}$:

$$\mathbf{S}_{bu} = \mathcal{N}\left(\mathbf{F}_{ct}^s + \mathbf{F}_{int}^s + \frac{1}{2}(\mathbf{F}_{rg}^s + \mathbf{F}_{by}^s) + \frac{1}{4}\sum_\theta \mathbf{F}_{o_\theta}^s\right) \tag{3}$$

Given the following two assumptions: 1) the selection process guided by the space-based bottom-up attention is a random event, and 2) the sample space of this random event is composed of all spatial locations in the image, the salience of a spatial location can be used to represent the degree of belief that bottom-up attention selects that location. Therefore, the probability of a spatial location $\mathbf{r}_i$ being attended by the bottom-up attention mechanism can be estimated as:

$$p_{bu}(\mathbf{r}_i) = \frac{S_{bu}(\mathbf{r}_i)}{\sum_{\mathbf{r}_{i'}} S_{bu}(\mathbf{r}_{i'})}, \tag{4}$$

where $p_{bu}(\mathbf{r}_i)$ denotes the probability of a spatial location $\mathbf{r}_i$ being attended by the bottom-up attention, and the denominator $\sum_{\mathbf{r}_{i'}} S_{bu}(\mathbf{r}_{i'})$ is the normalizing constant.

5.2 Top-down attention

5.2.1 LTM task representations and task-relevant objects

The *task-relevant object* can be defined as an object whose occurrence is expected by the task. Consistent with the autonomous mental development (AMD) paradigm (Weng et al., 2001), this paper proposes that actions include external actions that operate effectors and internal actions that predict the next possible attentional state (i.e., attentional prediction). Since the proposed perception system is object-based, the attentional prediction can be seen as the task-relevant object. Thus this paper models the *LTM task representation* as the association between attentional states and attentional prediction and uses it to deduce the task-relevant object.

It can be further proposed that the LTM task representation can be modeled by using a first-order discrete Markov process (FDMP). The FDMP can be expressed as $p(a_{t+1}|a_t)$, where a_t denotes the attentional state at time t and a_{t+1} denotes the attentional prediction for time $t+1$. This definition means that the probability of each attentional prediction for the next time can be estimated given the attentional state at the current time. The discrete attentional states is composed of LTM object representations.

5.2.2 Task-relevant feature

According to the IC hypothesis, it is required to deduce the task-relevant feature from the task-relevant object. This paper defines the *task-relevant feature* as a property that can discriminate the object from others. Although several autonomous factors (e.g., rewards obtained from learning) could be used, this paper uses the *conspicuity* quantity since it is one of the important intrinsic and innate properties of an object for measuring the discriminability. Through a training process that statistically encapsulates the conspicuity quantities obtained under different viewing conditions, a *salience descriptor* is achieved in the LTM object representation (See details in section 6.2 and section 6.3).

Therefore the salience descriptor is used to deduce the task-relevant feature by finding the feature dimension that has the greatest conspicuity. This deduction can be expressed as:

$$(f_{rel}, j_{rel}) = \arg \max_{f \in \{ct,int,rg,by,o_\theta\}} \max_{j \in \{1,2,...,N_j\}} \frac{\bar{\mu}_f^{s,j}}{1 + \bar{\sigma}_f^{s,j}}, \tag{5}$$

where N_j denotes the number of parts when $f \in \{int, rg, by, o_\theta\}$ and $N_j = 1$ when $f = ct$, $\bar{\mu}_f^{s,j}$ and $\bar{\sigma}_f^{s,j}$ respectively denote the mean and STD of salience descriptors in terms of a feature f in the LTM representation of the task-relevant object, f_{rel} denotes the *task-relevant feature dimension*, and j_{rel} denotes the index of the *task-relevant part*. The LTM object representation can be seen in section 6.3.

In the proposed system, the most task-relevant feature is first selected for guiding top-down attention. If the post-attentive recognition shows that the attended object is not the target, then the next task-relevant feature is joined. This process does not stop until the attended object is verified or all features are used.

5.2.3 Attentional template

Given the task-relevant feature dimension, its *appearance descriptor* in the LTM representation of the task-relevant object is used to build an attentional template in WM so as to estimate top-down biases. The attentional template is denoted as $\mathbf{F}_f^t$, where $f \in \{ct, int, rg, by, o_\theta\}$. The appearance descriptor will be presented in section 6.3.

5.2.4 Estimation of top-down biases

Bayesian inference is used to estimate the location-based top-down bias, which represents the probability of a spatial location being an instance of the task-relevant object. It can be generally expressed as:

$$p_{td}(\mathbf{r}_i|\mathbf{F}_f^t) = \frac{p_{td}(\mathbf{F}_f^t|\mathbf{r}_i) \times p_{td}(\mathbf{r}_i)}{\sum_{\mathbf{r}_{i'}} p_{td}(\mathbf{F}_f^t|\mathbf{r}_{i'}) \times p_{td}(\mathbf{r}_{i'})}, \tag{6}$$

where $p_{td}(\mathbf{r}_i)$ denotes the prior probability of a location $\mathbf{r}_i$ being attended by the top-down attention, $p_{td}(\mathbf{F}_f^t|\mathbf{r}_i)$ denotes the observation likelihood, $p_{td}(\mathbf{r}_i|\mathbf{F}_f^t)$ is the posterior probability of the location $\mathbf{r}_i$ being attended by the top-down attention given the attentional template $\mathbf{F}_f^t$. Assuming that the prior probability $p_{td}(\mathbf{r}_i)$ is a uniform distribution, Eq. (6) can be simplified into estimating the observation likelihood $p_{td}(\mathbf{F}_f^t|\mathbf{r}_i)$. The detailed estimation of $p_{td}(\mathbf{F}_f^t|\mathbf{r}_i)$ for each feature dimension, including contour, intensity, red-green, blue-yellow and orientations can be seen in our previous object-based visual attention (OVA) model (Yu et al., 2010).

5.2.5 Discussion

Compared with existing top-down attention methods, e.g., (Navalpakkam & Itti, 2005; Rao & Ballard, 1995a), the proposed method has four advantages. The first advantage is effectiveness due to the use of both salience and appearance descriptors. These two descriptors reciprocally aid each other: The salience descriptor guarantees that the task-relevant object can be effectively discriminated from distracters in terms of appearance, while the appearance descriptor can deal with the case that the task-relevant object and distracters have similar task-relevance values but different appearance values. The second advantage is efficiency. The computational complexity of (Rao & Ballard, 1995a) and our method can be approximated as $\mathcal{O}(d_h)$ and $\mathcal{O}(d_f^{few} d_l)$ respectively, where d_h denotes the dimension number of a high-level object representation, e.g., iconic representation (Rao & Ballard, 1995b) used in (Rao & Ballard, 1995a), d_l denotes the dimension number of a pre-attentive feature and d_f^{few} denotes the number of one or a few pre-attentive features used in our method. Since $d_h \gg d_f^{few} d_l$, the computation of our method is much cheaper. The third advantage is adaptability. As shown in (5), the task-relevant feature(s) can be autonomously deduced from the learned LTM representation such that the requirement of redesigning the representation of the task-relevant object for different tasks is eliminated. The fourth advantage is robustness. As shown in (6), the proposed method gives a bias toward the task-relevant object by using Bayes' rule, such that it is robust to work with noise, occlusion and a variety of viewpoints and illuminative effects.

5.3 Combination of bottom-up saliency and top-down biases

Assuming that bottom-up attention and top-down attention are two random events that are independent, the probability of an item being attended can be modeled as the probability of occurrence of either of these two events on that item. Thus, the probabilistic location-based attentional activation, denoted as $p_{attn}(\mathbf{r}_i)$, can be obtained by combining bottom-up saliency and top-down biases:

$$
\begin{cases}
p_{attn}(\mathbf{r}_i) = p_{bu}(\mathbf{r}_i) + p_{td}(\mathbf{r}_i | \{\mathbf{F}_f^t\}) - p_{bu}(\mathbf{r}_i) \times p_{td}(\mathbf{r}_i | \{\mathbf{F}_f^t\}) & \text{if } w_{bu} = 1 \text{ and } w_{td} = 1 \\
p_{attn}(\mathbf{r}_i) = p_{bu}(\mathbf{r}_i) & \text{if } w_{bu} = 1 \text{ and } w_{td} = 0, \quad (7) \\
p_{attn}(\mathbf{r}_i) = p_{td}(\mathbf{r}_i | \{\mathbf{F}_f^t\}) & \text{if } w_{bu} = 0 \text{ and } w_{td} = 1
\end{cases}
$$

where w_{bu} and w_{td} are two logic variables used as the conscious gating for bottom-up attention and top-down attention respectively and these two variables are set according to the task.

5.4 Proto-object based attentional activation

According to the IC hypothesis, it can be seen that a competitive advantage over an object is produced by directing attention to a spatial location in that object. Thus the probability of a proto-object being attended can be calculated using the *logic or* operator on the location-based probabilities. Furthermore, it can be assumed that two locations being attended are mutually exclusive according to the space-based attention theory (Posner et al., 1980). As a result, the probability of a proto-object $\mathbf{R}_g$ being attended, denoted as $p_{attn}(\mathbf{R}_g)$, can be calculated as:

$$
p_{attn}(\mathbf{R}_g) = \frac{1}{N_g} \sum_{\mathbf{r}_i \in \mathbf{R}_g} p_{td}(\mathbf{r}_i | \mathbf{F}_f^t), \tag{8}
$$

where $\mathbf{R}_g$ denotes a proto-object, N_g denotes the number of pixels in $\mathbf{R}_g$. The inclusion of $1/N_g$ is to eliminate the influence of the proto-object's size. The FOA is directed to the proto-object with maximal attentional activation.

6. Post-attentive perception

The flow chart of the post-attentive perception can be illustrated in Fig. 5. Four modules, as presented in section 3, are **interactive** during this stage.

6.1 Perceptual completion processing

This module works around the attended proto-object, denoted as $\mathbf{R}_{attn}^1$, to achieve the complete object region. It consists of two steps. The first step is recognition of the attended proto-object. This step explores LTM object representations in order to determine to which LTM object representation the attended proto-object belongs by using the post-attentive features. The extraction of post-attentive features and the recognition algorithm will be presented in section 6.2 and section 6.4 respectively. The matched LTM object representation, denoted as $\mathbf{O}_{attn}$, is then recalled from LTM.

The second step is completion processing:

1. If the local coding of $\mathbf{O}_{attn}$ includes multiple parts, several candidate proto-objects, which are spatially close to $\mathbf{R}_{attn}^1$, are selected from the current scene. They are termed as *neighbors* and denoted as a set $\{\mathbf{R}_n\}$.

2. The local post-attentive features are extracted in each $\mathbf{R}_n$.

3. Each $\mathbf{R}_n$ is recognized using the local post-attentive features and the matched LTM object representation $\mathbf{O}_{attn}$. If it is recognized as a part of $\mathbf{O}_{attn}$, it will be labeled as a part of the attended object. Otherwise, it will be eliminated.

4. Continue *item 2* and *item 3* iteratively until all neighbors have been checked.

These labeled proto-objects constitute the complete region of the attended object, which is denoted as a set $\{\mathbf{R}_{attn}\}$.

6.2 Extraction of post-attentive features

Post-attentive features $\mathbf{\breve{F}}$ are estimated by using the statistics within the attended object. They consist of *global post-attentive features* $\mathbf{\breve{F}}_{gb}$ and *local post-attentive features* $\mathbf{\breve{F}}_{lc}$. Each $\mathbf{\breve{F}}$ consists of *appearance component* $\mathbf{\breve{F}}^a$ and *salience component* $\mathbf{\breve{F}}^s$.

6.2.1 Local post-attentive features

Each proto-object, denoted as $\mathbf{R}_{attn}^m$, in the complete region being attended (i.e., $\mathbf{R}_{attn}^m \in \{\mathbf{R}_{attn}\}$) is the unit for estimating local post-attentive features. They can be estimated as a set that can be expressed as: $\{\mathbf{\breve{F}}_{lc}\} = \left\{ \left(\mathbf{\breve{F}}_{lc}^a(\mathbf{R}_{attn}^m), \mathbf{\breve{F}}_{lc}^s(\mathbf{R}_{attn}^m) \right)^T \right\}_{\forall \mathbf{R}_{attn}^m \in \{\mathbf{R}_{attn}\}}$.

The appearance components in an entry $\mathbf{\breve{F}}_{lc}$, denoted as $\mathbf{\breve{F}}_{lc}^a = \{\mathbf{\breve{F}}_f^a\}$ with $f \in \{int, rg, by, o_\theta\}$, are estimated by using the mean $\tilde{\mu}_f^{a,m}$ of $\mathbf{R}_{attn}^m$ in terms of f, i.e., $\mathbf{\breve{F}}_f^a(\mathbf{R}_{attn}^m) = \tilde{\mu}_f^{a,m}$.

The salience components, denoted as $\mathbf{\breve{F}}_{lc}^s = \{\mathbf{\breve{F}}_f^s\}$ with $f \in \{int, rg, by, o_\theta\}$, can be estimated using the mean of conspicuity $\tilde{\mu}_f^{s,m}$ of a $\mathbf{R}_{attn}^m$ in terms of f, i.e., $\mathbf{\breve{F}}_f^s(\mathbf{R}_{attn}^m) = \tilde{\mu}_f^{s,m}$. The conspicuity quantity F_f^s in terms of f is calculated using (2).

6.2.2 Global post-attentive features

The global post-attentive feature $\mathbf{\breve{F}}_{gb}$ is estimated after the complete region of the attended object, i.e., $\{\mathbf{R}_{attn}\}$, is obtained. Since the active contour technique (Blake & Isard, 1998; MacCormick, 2000) is used to represent a contour in this paper, the estimation of $\mathbf{\breve{F}}_{gb}$ includes

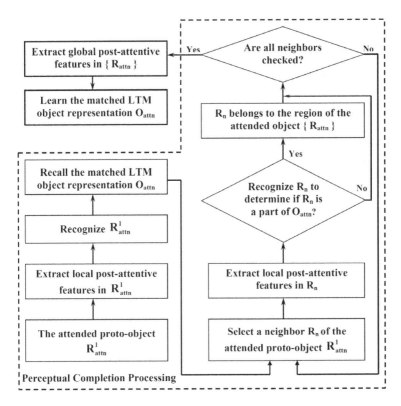

Fig. 5. The flowchart of the post-attentive perception stage.

two steps. The first step is to extract control points, denoted as a set $\{r_{cp}\}$, of the attended object's contour by using the method in our previous work (Yu et al., 2010). That is, each control point is an entry in the set $\{\tilde{\mathbf{F}}_{gb}\}$. The second step is to estimate the appearance and salience components at these control points, i.e., $\{\tilde{\mathbf{F}}_{gb}\} = \left\{\left(\tilde{\mathbf{F}}_{gb}^{a}(\mathbf{r}_{cp}), \tilde{\mathbf{F}}_{gb}^{s}(\mathbf{r}_{cp})\right)^{T}\right\}_{\forall \mathbf{r}_{cp}}$. The appearance component of an entry consists of spatial coordinates in the reference frame at a control point, i.e., $\tilde{\mathbf{F}}_{gb}^{a}(\mathbf{r}_{cp}) = \left(x_{\mathbf{r}_{cp}}\ y_{\mathbf{r}_{cp}}\right)^{T}$. The salience component of an entry is built by using the conspicuity value $F_{ct}^{s}(\mathbf{r}_{cp})$ in terms of pre-attentive contour feature at a control point, i.e., $\tilde{\mathbf{F}}_{gb}^{s}(\mathbf{r}_{cp}) = F_{ct}^{s}(\mathbf{r}_{cp})$.

6.3 Development of LTM object representations

The LTM object representation also consists of the local coding (denoted as $\mathbf{O}_{lc}$) and global coding (denoted as $\mathbf{O}_{gb}$). Each coding also consists of *appearance descriptors* (denoted as $\mathbf{O}^{a}$) and *salience descriptors* (denoted as $\mathbf{O}^{s}$). The PNN (Specht, 1990) is used to build them.

6.3.1 PNN of local coding

The PNN of a local coding $\mathbf{O}_{lc}$ (termed as a *local PNN*) includes three layers. The input layer receives the local post-attentive feature vector $\tilde{\mathbf{F}}_{lc}$. Each radial basis function (RBF) at the hidden layer represents a part of the learned object and thereby this layer is called a *part layer*. The output layer is a probabilistic mixture of all parts belonging to the object and thereby this layer is called an *object layer*.

The probability distribution of a RBF at the part layer of the local PNN can be expressed as:

$$p_j^k(\tilde{\mathbf{F}}_{lc}) = \mathcal{G}(\tilde{\mathbf{F}}_{lc}; \boldsymbol{\mu}_j^k, \boldsymbol{\Sigma}_j^k)$$

$$= \frac{1}{(2\pi)^{\frac{d}{2}} |\boldsymbol{\Sigma}_j^k|^{\frac{1}{2}}} \exp\{-\frac{1}{2}(\tilde{\mathbf{F}}_{lc} - \boldsymbol{\mu}_j^k)^T (\boldsymbol{\Sigma}_j^k)^{-1} (\tilde{\mathbf{F}}_{lc} - \boldsymbol{\mu}_j^k)\} \tag{9}$$

where $\mathcal{G}$ denotes the Gaussian distribution, $\boldsymbol{\mu}_j^k$ and $\boldsymbol{\Sigma}_j^k$ denote the mean vector and covariance matrix of a RBF, j is the index of a part, k is the index of an object in LTM, and d is the dimension number of a local post-attentive feature $\tilde{\mathbf{F}}_{lc}$. Since all feature dimensions are assumed to be independent, $\boldsymbol{\Sigma}_j^k$ is a diagonal matrix and standard deviation (STD) values of all feature dimensions of a RBF can constitute an STD vector $\boldsymbol{\sigma}_j^k$.

The probabilistic mixture estimation $r^k(\tilde{\mathbf{F}}_{lc})$ at the object layer can be expressed as:

$$r^k(\tilde{\mathbf{F}}_{lc}) = \sum_i \pi_j^k p_j^k(\tilde{\mathbf{F}}_{lc}), \tag{10}$$

where π_j^k denotes the contribution of part j to object k, which holds $\sum_j \pi_j^k = 1$.

6.3.2 PNN of global coding

The PNN for a global coding $\mathbf{O}_{gb}$ (termed as a *global PNN*) also includes three layers. The input layer receives the global post-attentive feature vector $\tilde{\mathbf{F}}_{gb}$. Each node of the hidden layer is a control point along the contour and thereby this layer is called a *control point layer*. The output layer is a probabilistic combination of all control points belonging to the object and thereby this layer is called an *object layer*. The mathematical expression of the global PNN is similar to the local PNN.

6.3.3 Learning of LTM object representations

Since the number of nodes (i.e., the numbers of parts and control points) is unknown and might be dynamically changed during the training course, this paper proposes a dynamical learning algorithm by using both the maximum likelihood estimation (MLE) and a Bayes' classifier to update the local and global PNNs at each time. This proposed dynamical learning algorithm can be summarized as follows. The Bayes' classifier is used to classify the training pattern to an existing LTM pattern. If the training pattern can be classified to an existing LTM pattern at the part level in a local PNN or at the control point level in a global PNN, both appearance and salience descriptors of this existing LTM pattern are updated using MLE. Otherwise, a new LTM pattern is created. Two thresholds τ_1 and τ_2 are introduced to determine the minimum correct classification probability to an existing part and an existing control point respectively. Algorithm 1 shows the learning routine of global and local codings. In the algorithm, a_j^k denotes the occurrence number of an existing pattern indexed by j of object k and it is initialized by 0, N_k denotes the number of parts in the local PNN or control points in the global PNN of object k, $.^2$ denotes the element-by-element square operator, and σ_{init} is a predefined STD value when a new pattern is created.

Algorithm 1 Learning Routine of Local and Global Codings

1: Given a local or global training pattern $(\tilde{\mathbf{F}}_{lc}, k)$ or $(\tilde{\mathbf{F}}_{gb}, k)$:
2: Set $\tilde{\mathbf{F}} = \tilde{\mathbf{F}}_{lc}$ or $\tilde{\mathbf{F}} = \tilde{\mathbf{F}}_{gb}$;
3: Recognize $\tilde{\mathbf{F}}$ to obtain a recognition probability $p_i^k(\tilde{\mathbf{F}})$;
4: **if** $p_j^k(\tilde{\mathbf{F}}) \geq \tau_1$ or $\geq \tau_2$ **then**
5:     // *Update part j of object k*
6:     $\sigma_{temp} = [a_j^k(\sigma_j^k).^2 + a_j^k(\mu_j^k).^2 + (\tilde{\mathbf{F}}).^2]/(a_j^k + 1);$ // *Prepare for updating the STD*
7:     $\mu_j^k = (a_j^k \mu_j^k + \tilde{\mathbf{F}})/(a_j^k + 1);$ // *Update the mean vector*
8:     $\sigma_j^k = [\sigma_{temp}^d - (\mu_j^k).^2].^{-\frac{1}{2}};$ // *Update the STD*
9:     $a_j^k = a_j^k + 1;$ // *Increment the occurrence number*
10: **else**
11:     // *Create a new part i of object k*
12:     Set $N_k = N_k + 1;$ $i = N_k;$
13:     $\mu_j^k = \tilde{\mathbf{F}};$ $\sigma_j^k = \sigma_{init};$ $a_j^k = 1;$ // *Set the initial mean, STD and occurrence number*
14: **end if**
15: $\forall j: \pi_j^k = a_j^k / \sum_{j'} a_{j'}^k.$ // *Normalize weights π*

6.4 Object recognition

Due to the page limitation, the object recognition module can be summarized as follows. It can be modeled at two levels. The first one is the object level. The purpose of this level is to recognize to which LTM object an attended pattern belongs. The second one is the part level or control point level. Recognition at this level is performed given an LTM object to which the attended pattern belongs. Thus, the purpose of this level is to recognize to which part in a local PNN or to which control point in a global PNN an attended pattern belongs. At each level, object recognition can generally be modeled as a decision unit by using Bayes' theorem. Assuming that the prior probability is equal for all LTM patterns at each level, the observation likelihood can be seen as the posterior probability.

7. Experiments

This proposed autonomous visual perception system is tested in the task of object detection. The unconscious perception path (i.e., the bottom-up attention module) can be used to detect a salient object, such as a landmark, whereas the conscious perception path (i.e., the top-down attention module) can be used to detect the task-relevant object, i.e., the expected target. Thus the unconscious and conscious aspects are tested in two robotics tasks respectively: One is detecting a salient object and the other is detecting a task-relevant object.

7.1 Detecting a salient object

The salient object is an unusual or unexpected object and the current task has no prediction about its occurrence. There are three objectives in this task. The first objective is to illustrate the unconscious capability of the proposed perception system. The second objective is to show the advantages of using object-based visual attention for perception by comparing it with the space-based visual attention methods. The third objective is to show the advantage of integrating the contour feature into the bottom-up competition module. The result is that an object that has a conspicuous shape compared with its neighbors can be detected. Two experiments are shown in this section, including the detection of an object that is conspicuous in colors and in contour respectively.

7.1.1 Experimental setup

Artificial images are used in the experiments. The frame size of all images is 640×480 pixels. In order to show the robustness of the proposed perception system, these images are obtained using different settings, including noise, spatial transformation and changes of lighting. The noisy images are manually obtained by adding salt and pepper noise patches (noise density: $0.1 \sim 0.15$, patch size: 10×10 pixels $\sim 15 \times 15$ pixels) into original r, g and b color channels respectively. The experimental results are compared with the results of Itti's model (i.e., space-based bottom-up attention) (Itti et al., 1998) and Sun's model (i.e., object-based bottom-up attention) (Sun & Fisher, 2003).

7.1.2 An object conspicuous in colors

The first experiment is detecting an object that is conspicuous to its neighbors in terms of colors and all other features are approximately the same between the object and its neighbors. The experimental results are shown in Fig. 6. The salient object is the red ball in this experiment. Results of the proposed perception system are shown in Fig. 6(d), which indicate that this proposed perception system can detect the object that is conspicuous to its neighbors in terms of colors in different settings. Results of Itti's model and Sun's model are shown in Fig. 6(e) and Fig. 6(f) respectively. It can be seen that Itti's model fails to detect the salient object when noise is added to the image, as shown in column 2 in Fig. 6(e). This indicates that the proposed object-based visual perception system is more robust to noise than the space-based visual perception methods.

7.1.3 An object conspicuous in contour

The second experiment is detecting an object that is conspicuous to its neighbors in terms of contour and all other features are approximately the same between the object and its neighbors. The experimental results are shown in Fig. 7. In this experiment, the salient object is the triangle. Detection results of the proposed perception system are shown in Fig. 7(d), which indicate that the proposed perception system can detect the object that is conspicuous to its neighbors in terms of contour in different settings. Detection results of Itti's model and Sun's model are shown in Fig. 7(e) and Fig. 7(f) respectively. It can be seen that both Itti's model and Sun's model fail to detect the salient object when noise is added to the image, as shown in column 2 in Fig. 7(e) and Fig. 7(f) respectively. This experiment indicates that the proposed object-based visual perception system is capable of detecting the object conspicuous in terms of contour in different settings due to the inclusion of contour conspicuity in the proposed bottom-up attention module.

7.2 Detecting a task-relevant object

It is an important ability for robots to accurately detect a task-relevant object (i.e., target) in the cluttered environment. According to the proposed perception system, the detection procedure consists of two phases: a learning phase and a detection phase. The objective of the learning phase is to develop the LTM representation of the target. The objective of the detection phase is to detect the target by using the learned LTM representation of the target. The detection phase can be implemented as a two-stage process. The first stage is attentional selection: The

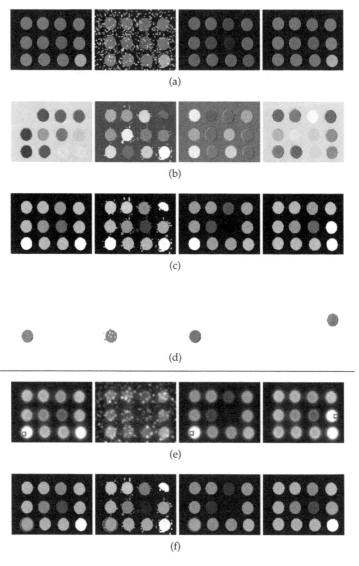

(a)

(b)

(c)

(d)

(e)

(f)

Fig. 6. Detection of a salient object, which is conspicuous to its neighbors in terms of colors. Each column represents a type of experimental setting. Column 1 is a typical setting. Column 2 is a noise setting of column 1. Column 3 is a different lighting setting with respect to column 1. Column 4 is a spatial transformation setting with respect to column 1. Row (a): Original input images. Row (b): Pre-attentive segmentation. Each color represents one proto-object. Row (c): Proto-object based attentional activation map. Row (d): The complete region being attended. Row (e): Detection results using Itti's model. The red rectangles highlight the attended location. Row (f): Detection results using Sun's model. The red circles highlight the attended object.

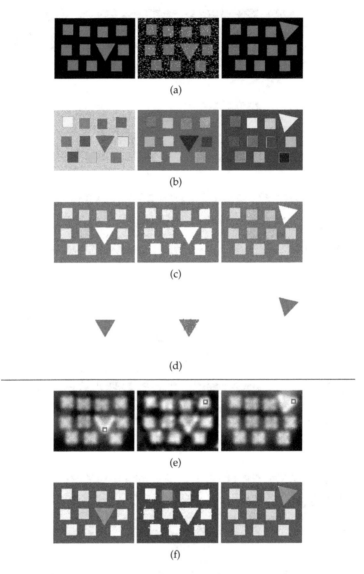

Fig. 7. Detection of a salient object, which is conspicuous to its neighbors in terms of contour. Each column represents a type of experimental setting. Column 1 is a typical setting. Column 2 is a noise setting of column 1. Column 3 is a spatial transformation setting with respect to column 1. Row (a): Original input images. Row (b): Pre-attentive segmentation. Each color represents one proto-object. Row (c): Proto-object based attentional activation map. Row (d): The complete region being attended. Row (e): Detection results using Itti's model. The red rectangles highlight the attended location. Row (f): Detection results using Sun's model. The red circles highlight the attended object.

task-relevant feature(s) of the target is used to guide attentional selection through top-down biasing to obtain an attended object. The second stage is post-attentive recognition: The attended object is recognized using the target's LTM representation to check if it is the target. If not, another procedure of attentional selection is performed by using more task-relevant features.

7.2.1 Experimental setup

Two objects are used to test the proposed method of detecting a task-relevant object: a book and a human. Images and videos are obtained under different settings, including noise, transformation, lighting changes and occlusion. For training for the book, 20 images are used. For testing for the book, 50 images are used. The size of each image is 640×480 pixels. For detecting the human, three videos are obtained by a moving robot. Two different office environments have been used. Video 1 and video 2 are obtained in office scene 1 with low and high lighting conditions respectively. Video 3 is obtained in office scene 2. All three videos contain a total of 650 image frames, in which 20 image frames are selected from video 1 and video 2 for training and the rest of the 630 image frames are used for testing. The size of each frame in these videos is 1024×768 pixels. It is important to note that each test image includes not only a target but also various distracters. The noisy images are manually obtained by adding salt and pepper noise patches (noise density: 0.1, patch size: 5×5 pixels) into original r, g and b color channels respectively.

The results of the proposed method are compared with the results of Itti's model (Itti et al., 1998) (i.e., a space-based bottom-up attention model) and Navalpakkam's model (Navalpakkam & Itti, 2005) (i.e., a space-based top-down attention model) respectively.

7.2.2 Task 1

The first task is to detect the book that has multiple parts. The learned LTM representation of the book is shown in Table 1, which has shown that the book has two parts and the blue-yellow feature in the first part can be deduced as the task-relevant feature dimension since the value $\mu^s/(1+\sigma^s)$ of this feature is maximal. Detection results of the proposed perception system are shown in Fig. 8(d). It can be seen that the book is successfully detected. Results of Itti's model and Navalpakkam's model, as shown in Fig. 8(e) and Fig. 8(f) respectively, show that these models fail to detect the target in some cases.

7.2.3 Task 2

The second task is to detect a human. Table 2 has shown that the human has two parts (including face and body) and the contour feature can be deduced as the task-relevant feature dimension since the value $\mu^s/(1+\sigma^s)$ of this feature is maximal. Detection results of the proposed perception system are shown in Fig. 9(d). It can be seen that the human is successfully detected. Results of Itti's model and Navalpakkam's model, as shown in Fig. 9(e) and Fig. 9(f) respectively, show that these models fail to detect the target in most cases.

7.2.4 Performance evaluation

Performance of detecting task-relevant objects is evaluated using true positive rate (TPR) and false positive rate (FPR), which are calculated as:

$$TPR = TP/nP, \tag{11}$$

f	j	μ^a	σ^a	μ^s	σ^s	$\mu^s/(1+\sigma^s)$
ct	1		-	75.0	19.7	3.6
int	1	106.6	5.8	27.9	14.5	1.8
rg	1	22.1	8.7	199.6	18.2	10.4
by	1	-108.0	9.1	215.6	8.7	**22.2**
$o_{0°}$	1	N/A	N/A	41.8	9.8	3.9
$o_{45°}$	1	N/A	N/A	41.4	12.8	3.0
$o_{90°}$	1	N/A	N/A	34.7	16.3	2.0
$o_{135°}$	1	N/A	N/A	46.5	15.7	2.8
int	2	60.5	8.2	80.0	5.7	11.9
rg	2	0.4	4.3	18.3	6.4	2.5
by	2	120.8	6.7	194.7	8.1	21.4
$o_{0°}$	2	N/A	N/A	48.5	11.1	4.0
$o_{45°}$	2	N/A	N/A	53.8	9.9	4.9
$o_{90°}$	2	N/A	N/A	38.4	14.6	2.5
$o_{135°}$	2	N/A	N/A	59.4	20.3	2.8

Table 1. Learned LTM object representation of the book. f denotes a pre-attentive feature dimension. j denotes the index of a part. The definitions of μ^a, σ^a, μ^s and σ^s can be seen in section 5.2.2.

f	j	μ^a	σ^a	μ^s	σ^s	$\mu^s/(1+\sigma^s)$
ct	1		-	68.3	6.9	**8.6**
int	1	28.4	21.7	18.8	13.9	1.3
rg	1	-7.0	7.1	28.6	10.8	2.4
by	1	10.9	5.4	48.4	10.9	4.1
$o_{0°}$	1	N/A	N/A	33.4	6.7	4.3
$o_{45°}$	1	N/A	N/A	39.8	11.4	3.2
$o_{90°}$	1	N/A	N/A	37.4	6.1	5.3
$o_{135°}$	1	N/A	N/A	37.5	13.5	2.6
int	2	52.0	12.5	25.6	15.6	1.5
rg	2	-2.3	17.4	49.5	18.8	2.5
by	2	-29.3	6.9	60.4	22.3	2.6
$o_{0°}$	2	N/A	N/A	12.1	6.6	1.6
$o_{45°}$	2	N/A	N/A	16.5	8.3	1.8
$o_{90°}$	2	N/A	N/A	15.0	7.9	1.7
$o_{135°}$	2	N/A	N/A	17.2	8.1	1.9

Table 2. Learned LTM object representation of the human. f denotes a pre-attentive feature dimension. j denotes the index of a part. The definitions of μ^a, σ^a, μ^s and σ^s can be seen in section 5.2.2.

$$FPR = FP/nN, \qquad (12)$$

where nP and nN are numbers of positive and negative objects respectively in the testing image set, TP and FP are numbers of true positives and false positives. The positive object is the target to be detected and the negative objects are distracters in the scene.

Detection performance of the proposed perception system and other visual attention based methods is shown in Table 3. Note that "Naval's" represents Navalpakkam's method.

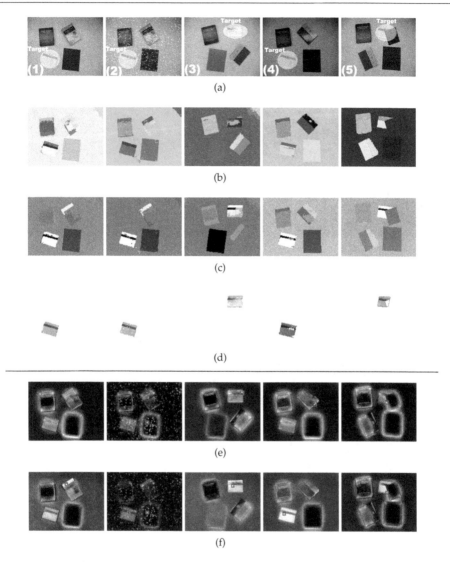

Fig. 8. Detection of the book. Each column represents a type of experimental setting. Column 1 is a typical setting. Column 2 is a noise setting of column 1. Column 3 is a spatial transformation (including translation and rotation) setting with respect to column 1. Column 4 is a different lighting setting with respect to column 1. Column 5 is an occlusion setting. Row (a): Original input images. Row (b): Pre-attentive segmentation. Each color represents one proto-object. Row (c): Proto-object based attentional activation map. Brightness represents the attentional activation value. Row (d): The complete region of the target. The red contour in the occlusion case represents the illusory contour (Lee & Nguyen, 2001), which shows the post-attentive perceptual completion effect. Row (e): Detection results using Itti's model. The red rectangle highlights the most salient location. Row (f): Detection results using Navalpakkam's model. The red rectangle highlights the most salient location.

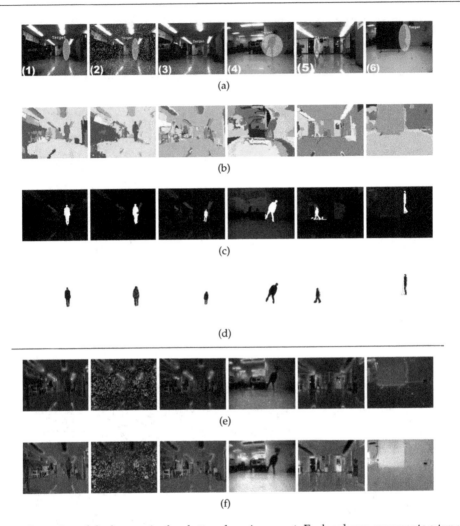

Fig. 9. Detection of the human in the cluttered environment. Each column represents a type of experimental setting. Column 1 is a typical setting (from video 1). Column 2 is a noise setting of column 1. Column 3 is a scaling setting with respect to column 1 (from video 1). Column 4 is a rotation setting with respect to column 1 (from video 3). Column 5 is a different lighting setting with respect to column 1 (from video 2). Column 6 is an occlusion setting (from video 3). Row (a): Original input images. Row (b): Pre-attentive segmentation. Each color represents one proto-object. Row (c): Proto-object based attentional activation map. Brightness represents the attentional activation value. Row (d): The complete region of the target. The red contour in the occlusion case represents the illusory contour (Lee & Nguyen, 2001), which shows the post-attentive perceptual completion effect. Row (e): Detection results using Itti's model. The red rectangle highlights the most salient location. Row (f): Detection results using Navalpakkam's model. The red rectangle highlights the most salient location.

Task	Method	TP	FP	nP	nN	TPR (%)	FPR (%)
1	Proposed	47	3	50	244	94.00	1.23
	Itti's	16	34	50	244	32.00	13.93
	Naval's	41	9	50	244	82.00	3.69
2	Proposed	581	49	630	30949	92.22	0.16
	Itti's	5	625	630	30949	0.79	2.02
	Naval's	36	594	630	30949	5.71	1.92

Table 3. Performance of detecting task-relevant objects.

8. Conclusion

This paper has presented an autonomous visual perception system for robots using the object-based visual attention mechanism. This perception system provides the following four contributions. The first contribution is that the attentional selection stage supplies robots with the cognitive capability of knowing how to perceive the environment according to the current task and situation, such that this perception system is adaptive and general to any task and environment. The second contribution is the top-down attention method using the IC hypothesis. Since the task-relevant feature(s) are conspicuous, low-level and statistical, this top-down biasing method is more effective, efficient and robust than other methods. The third contribution is the PNN based LTM object representation. This LTM object representation can probabilistically embody various instances of that object, such that it is robust and discriminative for top-down attention and object recognition. The fourth contribution is the pre-attentive segmentation algorithm. This algorithm extends the irregular pyramid techniques by integrating a scale-invariant probabilistic similarity measure, a similarity-driven decimation method and a similarity-driven neighbor search method. It provides rapid and satisfactory results of pre-attentive segmentation for object-based visual attention. Based on these contributions, this perception system has been successfully tested in the robotic task of object detection under different experimental settings.

The future work includes the integration of the bottom-up attention in the temporal context and experiments of the combination of bottom-up and top-down attention.

9. References

Aziz, M. Z., Mertsching, B., Shafik, M. S. E.-N. & Stemmer, R. (2006). Evaluation of visual attention models for robots, *Proceedings of the 4th IEEE Conference on Computer Vision Systems*, p. 20.

Backer, G., Mertsching, B. & Bollmann, M. (2001). Data- and model-driven gaze control for an active-vision system, *IEEE Transactions on Pattern Analysis and Machine Intelligence* 23(12): 1415–1429.

Baluja, S. & Pomerleau, D. (1997). Dynamic relevance: Vision-based focus of attention using artificial neural networks, *Artificial Intelligence* 97: 381–395.

Belardinelli, A. & Pirri, F. (2006). A biologically plausible robot attention model, based on space and time, *Cognitive Processing* 7(Supplement 5): 11–14.

Belardinelli, A., Pirri, F. & Carbone, A. (2006). Robot task-driven attention, *Proceedings of the International Symposium on Practical Cognitive Agents and Robots*, pp. 117–128.

Bhattacharyya, A. (1943). On a measure of divergence between two statistical populations defined by their probability distributions, *Bulletin of the Calcutta Mathematical Society* 35: 99–109.

Blake, A. & Isard, M. (1998). *Active Contours: The Application of Techniques from Graphics, Vision, Control Theory and Statistics to Visual Tracking of Shapes in Motion*, Springer-Verlag New York, Secaucus, NJ.

Breazeal, C., Edsinger, A., Fitzpatrick, P. & Scassellati, B. (2001). Active vision for sociable robots, *IEEE Transactions on Systems, Man, and Cybernetics - Part A: Systems and Humans* 31(5): 443–453.

Burt, P. J. & Adelson, E. H. (1983). The laplacian pyramid as a compact image code, *IEEE Transactions on Communications* 31(4): 532–540.

Carbone, A., Finzi, A. & Orlandini, A. (2008). Model-based control architecture for attentive robots in rescue scenarios, *Autonomous Robots* 24(1): 87–120.

Carpenter, G. A. & Grossberg, S. (2003). Adaptive resonance theory, *in* M. A. Arbib (ed.), *The Handbook of Brain Theory and Neural Networks, Second Edition*, MIT Press, Cambridge, MA, pp. 87–90.

Chikkerur, S., Serre, T., Tan, C. & Poggio, T. (2010). What and where: A bayesian inference theory of attention, *Visual Research* 50(22): 2233–2247.

Desimone, R. & Duncan, J. (1995). Neural mechanisms of selective visual attention, *Annual Reviews of Neuroscience* 18: 193–222.

Duncan, J. (1984). Selective attention and the organization of visual information, *Journal of Experimental Psychology: General* 113(4): 501–517.

Duncan, J. (1998). Converging levels of analysis in the cognitive neuroscience of visual attention, *Philosophical Transactions of The Royal Society Lond B: Biological Sciences* 353(1373): 1307–1317.

Duncan, J., Humphreys, G. & Ward, R. (1997). Competitive brain activity in visual attention, *Current Opinion in Neurobiology* 7(2): 255–261.

Frintrop, S. (2005). *VOCUS: A visual attention system for object detection and goal-directed search*, PhD thesis, University of Bonn, Germany.

Frintrop, S. & Jensfelt, P. (2008). Attentional landmarks and active gaze control for visual slam, *IEEE Transactions on Robotics* 24(5): 1054–1065.

Frintrop, S. & Kessel, M. (2009). Most salient region tracking, *Proceedings of the IEEE International Conference on Robotics and Automation (ICRA)*, pp. 1869–1874.

Greenspan, A. G., Belongie, S., Goodman, R., Perona, P., Rakshit, S. & Anderson, C. H. (1994). Overcomplete steerable pyramid filters and rotation invariance, *Proceedings of IEEE Computer Vision and Pattern Recognition*, pp. 222–228.

Grossberg, S. (2005). Linking attention to learning, expectation, competition, and consciousness, *in* L. Itti, G. Rees & J. Tsotsos (eds), *Neurobiology of Attention*, Elsevier, San Diego, CA, pp. 652–662.

Grossberg, S. (2007). Consciousness clears the mind, *Neural Networks* 20: 1040–1053.

Hoya, T. (2004). Notions of intuition and attention modeled by a hierarchically arranged generalized regression neural network, *IEEE Transactions on Systems, Man, and Cybernetics - Part B: Cybernetics* 34(1): 200–209.

Itti, L. & Baldi, P. (2009). Bayesian surprise attracts human attention, *Visual Research* 49(10): 1295–1306.

Itti, L., Koch, C. & Niebur, E. (1998). A model of saliency-based visual attention for rapid scene analysis, *IEEE Transactions on Pattern Analysis and Machine Intelligence* 20(11): 1254–1259.

Jolion, J. M. (2003). Stochastic pyramid revisited, *Pattern Recognition Letters* 24(8): 1035–1042.

Lee, T. S. & Nguyen, M. (2001). Dynamics of subjective contour formation in the early visual cortex, *Proceedings of the Natual Academy of Sciences of the United States of America* 98(4): 1907–1911.

MacCormick, J. (2000). *Probabilistic modelling and stochastic algorithms for visual localisation and tracking*, PhD thesis, Department of Engineering Science, University of Oxford.

Maier, W. & Steinbach, E. (2010). A probabilistic appearance representation and its application to surprise detection in cognitive robots, *IEEE Transactions on Autonomous Mental Development* 2(4): 267–281.

Meer, P. (1989). Stochastic image pyramids, *Computer Vision, Graphics, and Image Processing* 45(3): 269–294.

Montanvert, A., Meer, P. & Rosenfeld, A. (1991). Hierarchical image analysis using irregular tessellations, *IEEE Transactions on Pattern Analysis and Machine Intelligence* 13(4): 307–316.

Navalpakkam, V. & Itti, L. (2005). Modeling the influence of task on attention, *Vision Research* 45(2): 205–231.

Orabona, F., Metta, G. & Sandini, G. (2005). Object-based visual attention: a model for a behaving robot, *Proceedings of the IEEE Computer Society Conference on Computer Vision and Patter Recognition (CVPR)*, p. 89.

Posner, M. I., Snyder, C. R. R. & Davidson, B. J. (1980). Attention and the detection of signals, *Journal of Experimental Psychology: General* 14(2): 160–174.

Rao, R. P. N. & Ballard, D. H. (1995a). An active vision architecture based on iconic representations, *Artificial Intelligence* 78: 461–505.

Rao, R. P. N. & Ballard, D. H. (1995b). Object indexing using an iconic sparse distributed memory, *Proc. the 5th Intl. Conf. Computer Vision*, pp. 24–31.

Rao, R. P. N., Zelinsky, G. J., Hayhoe, M. M. & Ballard, D. H. (2002). Eye movements in iconic visual search, *Vision Research* 42: 1447–1463.

Scholl, B. J. (2001). Objects and attention: the state of the art, *Cognition* 80(1-2): 1–46.

Sharon, E., Brandt, A. & Basri, R. (2000). Fast multiscale image segmentation, *Proceedings of IEEE Conference on Computer Vision and Pattern Recognition (CVPR)*, pp. 70–77.

Sharon, E., Galun, M., Sharon, D., Basri, R. & Brandt, A. (2006). Hierarchy and adaptivity in segmenting visual scenes, *Nature* 442: 810–813.

Shi, J. & Malik, J. (2000). Normalized cuts and image segmentation, *IEEE Transactions on Pattern Analysis and Machine Intelligence* 22(8): 888–905.

Specht, D. F. (1990). Probabilistic neural networks, *Neural Networks* 3(1): 109–118.

Sun, Y. (2008). A computer vision model for visual-object-based attention and eye movements, *Computer Vision and Image Understanding* 112(2): 126–142.

Sun, Y. & Fisher, R. (2003). Object-based visual attention for computer vision, *Artificial Intelligence* 146(1): 77–123.

Tipper, S. P., Howard, L. A. & Houghton, G. (1998). Action-based mechanisms of attention, *Philosophical Transactions: Biological Sciences* 353(1373): 1385–1393.

Treisman, A. M. & Gelade, G. (1980). A feature integration theory of attention, *Cognition Psychology* 12(1–2): 507–545.

Tsotsos, J. K., Culhane, S. M., Wai, W. Y. K., Lai, Y., Davis, N. & Nuflo, F. (1995). Modelling visual attention via selective tuning, *Artificial Intelligence* 78: 282–299.

Walther, D., Rutishauser, U., Koch, C. & Perona, P. (2004). On the usefulness of attention for object recognition, *Workshop on Attention and Performance in Computational Vision at ECCV*, pp. 96–103.

Weng, J., McClelland, J., Pentland, A., Sporns, O., Stockman, I. & Thelen, E. (2001). Autonomous mental development by robots and animals, *Science* 291(5504): 599–600.

Wolfe, J. M. (1994). Guided search 2.0: A revised model of visual search, *Psychonomic Bulletin and Review* 1(2): 202–238.

Yu, Y., Mann, G. K. I. & Gosine, R. G. (2010). An object-based visual attention model for robotic applications, *IEEE Transactions on Systems, Man and Cybernetics, Part B: Cybernetics* 40(5): 1398–1412.

3D Visual Information for Dynamic Objects Detection and Tracking During Mobile Robot Navigation

D.-L. Almanza-Ojeda* and M.-A. Ibarra-Manzano
Digital Signal Processing Laboratory, Electronics Department; DICIS,
University of Guanajuato, Salamanca, Guanajuato
Mexico

1. Introduction

An autonomous mobile robot that navigates in outdoor environments requires functional and decisional routines enabling it to supervise the estimation and the performance of all its movements for carrying out an envisaged trajectory. At this end, a robot is usually equipped with several high-performance sensors. However, we are often interested in less complex and low-cost sensors that could provide enough information to detect in real-time when the trajectory is free of dynamic obstacles. In this context, our strategy was focused on visual sensors, particulary on stereo vision since this provides the depth coordinate for allowing a better perception of the environment. Visual perception for robot mobile navigation is a complex function that requires the presence of "salience" or "evident" patrons to identify something that "breaks" the continuous tendency of data. Usually, interesting points or segments are used for evaluating patrons in position, velocity, appearance or other characteristics that allows us forming groups (Lookingbill et al., 2007), (Talukder & Matthies, 2004). Whereas complete feature vectors are more expressive for explaining objects, here we use 3D feature points for proposing a strategy computationally less demanding conserving the main objective of the work: detect and track moving objects in real time.

This chapter presents a strategy for detecting and tracking dynamic objects using a stereo-vision system mounted on a mobile robot. First, a set of interesting points are extracted from the left image. A disparity map, provided by a real-time stereo vision algorithm implemented on FPGA, gives the 3D position of each point. In addition, velocity magnitude and orientation are obtained to characterize the set of points on the space R^6. Groups of dynamic 2D points are formed using the *a contrario* clustering technique in the 4D space and then evaluated on their depth value yielding groups of dynamic 3D-points. Each one of these groups is initialized by a convex contour with the velocity and orientation of the points given a first estimation of the dynamic object position and velocity. Then an active contour defines a more detailed silhouette of the object based on the intensity and depth value inside of the contour. It is well known that active contour techniques require a highly dense computations. Therefore, in order to reduce the time of processing a fixed number of iterations is used at each frame, so the convergence of the object real limits will be incrementally achieved along

---

*Part of this work was developed when authors were with LAAS-CNRS, Toulouse, France

several frames. A simple and predefined knowledge about the most usual dynamic objects found in urban environments are used to label the growing regions as a rigid or non-rigid object, essentially cars and people. Experiments on detection and tracking of vehicles and people, as well as during occlusion situations with a mobile robot in real world scenarios are presented and discussed.

2. Related works

The issue of moving object detection has been largely studied by the robotic and computer vision community. Proposed strategies use mainly a combination of active and passive sensors mounted on the mobile robot like laser (Vu & Aycard, 2009) with cameras (Katz et al., 2008), or infrared cameras (Matthies et al., 1998), just to name a few. However, a multi-sensor system requires to solve the problem of fusing the data from different sources which often requires more complex estimation cases. Indeed, more information could be acquired using several sensors but these systems are expensive and complex. To overcome those constraints, the proposed solution consist in using one or more cameras as the only source of information in the system (Talukder & Matthies, 2004), (Williamson, 1998). Essentially vision sensors provide enough information for localization and mapping (Sola et al., 2007) or for describing static and moving objects on the environment (Klappstein et al., 2008).

The stereo-vision is one of the most used techniques for reconstructing the 3D (depth) information of a scene from two images, called left and right. This information is acquired from two cameras separated by a previously established distance. The disparity map is a representation that contains the depth information of the scene. It is well known that dense stereo-vision delivers more complete information than sparse stereo-vision but this is a high-processing cost technique which enables to perform in real time using and ordinary computer system. We use a stereo-vision technique in order to detect moving objects but implemented on a re-configurable architecture that maximizes the efficiency of the system. In the last decade, several works have proposed the development of high-performance architectures to solve the stereo-vision problem i.e. digital signal processing (DSP), field programmable gate arrays (FPGA) or application-specific integrated circuits (ASIC). The ASIC devices are one of the most complicated and expensive solutions, however they afford the best conditions for developing a final commercial system (Woodfill et al., 2006). On the other hand, FPGA have allowed the creation of hardware designs in standard, high-volume parts, thereby amortizing the cost of mask sets and significantly reducing time-to-market for hardware solutions. However, engineering cost and design time for FPGA-based solutions still remain significantly higher than software-based solutions. Designers must frequently iterate the design process in order to achieve system performance requirements and simultaneously minimize the required size of the FPGA. Each iteration of this process takes hours or days to be completed (Schmit et al., 2000). Even if designing with FPGAs is faster than designing ASICs, it has a finite resource capacity which demands clever strategies for adapting versatile real-time systems (Masrani & MacLean, 2006).

2.1 Overall strategy for 3D dynamic object detection from a mobile robot

A seminal work of this strategy was presented in (Almanza-Ojeda et al., 2010) and (Almanza-Ojeda et al., 2011). Whereas the former proposes a monocamera strategy (Almanza-Ojeda et al., 2010), and the latter the fusion with the information provided by inertial (IMU) (Almanza-Ojeda et al., 2011), here we propose an extension of this strategy to a stereo vision images provided by a bank stereo mounted on the robot. The stereo images are processed by a stereo vision algorithm designed on FPGA that calculates the disparity map which provides the depth at each point of the input image. Thus, we use a combination of the 3D and 2D representation of feature points for grouping them according to similar distribution function in position and velocity. Moreover, these groups of points permit us to initialize an active contour for obtaining the object boundary. Therefore, the contour initialization of the detected points is fundamental in order to properly perform the shape recovering.

A block diagram of the proposed generalized algorithm is depicted in figure 1. A stereo vision module for FPGA based on Census Transform provides the disparity map using left and right images. At the same time, a selection-tracking process of feature points is carried out on a loop. However, rather than considering both stereo vision images during interest point selection, we will consider only left image for obtaining these 2D feature points. Further, each feature point will be associated with a depth value given by the disparity map. At this point it is necessary to point out that clustering task is performed until a small number of images have been processed. According to this, clustering module will receive a spatial-temporal set of points since each feature location and velocity have been accumulated through the time. The clustering method that we use is the *a contrario* method proposed by Veit et. al. (Veit et al., 2007). This clustering method consist in grouping the interesting points which have a "coherent" movement along a short number of consecutive images. Here the term coherent refers to movements that follow a similar and constant magnitude and direction described by the probability density function of the points under evaluation.

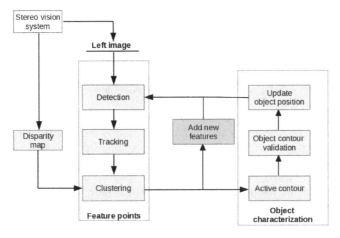

Fig. 1. Global strategy of object detection and identification using a stereo vision system mounted on a mobile robot

Once dynamic group of points have been detected by the clustering process, from this group of points it is possible to delimit a region and an irregular contour in the image with the aim of characterizing a probable dynamic object. Thus, by doing an analysis of intensity inside of this contour, we propose to match the provided region along the image sequence in order to follow the non-rigid objects that have more complicated movement, like pedestrians.

This work provides two distinct contributions: (1) the addition of depth information for performing the clustering of 3D set of dynamic points instead of only 2D as in the *a contrario* clustering technique (Veit et al., 2007), (2) the use of dynamic groups for initializing an irregular contour active that temporally recovers actual object boundary. The second contribution was preliminary discussed in (Almanza-Ojeda et al., 2011) where the same active contour was initialized for representing object region, however that only uses a snake based on contour information, so nothing is doing with features inside of the region.

The structure of the chapter is as follows. In section 3, we describe the stereo vision module and its performance. Then section 4.1 details interest 2D point selection and how the 3D points are obtained using the disparity map. The grouping technique of the interest points based on the *a contrario* clustering is explained in section 4.2. We present in section 5 object shape recovering using active contours. Section 6 contains our experimental results from indoor an outdoor environments. Finally, we end up with conclusions in section 7.

3. Stereo vision module

We use a disparity map calculated using the Census Transform algorithm (Zabih & Woodfill, 1994) implemented on a programmable device, in this case a FPGA (Field Programmable Gate Array). The architecture of the Census transform algorithm was developed by (Ibarra-Manzano, Devy, Boizard, Lacroix & Fourniols, 2009) in which left and right images acquired from the stereo vision bank are processed for generating up to 325 dense disparity maps of 640×480 pixels per second. It is important to point out that most of the vision-based systems do not require high video-frame rates because usually they are implemented on computers or embedded platforms which are not FPGA-based. As this is also our case, we have adapted the disparity map generation to the real-time application required by our system by tuning some configuration parameters in the architecture. A block diagram of the stereo vision algorithm is shown in figure 2. In the following, we will describe in general the calculation of the disparity map based on the Census transform. However, the architectural implementation on the FPGA is a problem that has not dealt with in this work, all these details will be found in (Ibarra-Manzano, Devy, Boizard, Lacroix & Fourniols, 2009), (Ibarra-Manzano & Almanza-Ojeda, 2011).

In the stereo vision algorithm each of the images (right and left) are processed independently in parallel. The process begins with the rectification and correction of the distortion for each image in order to decrease the size of the search of points to a single dimension during disparity calculation. This strategy is known as epipolar restriction in which, once the main axes of the cameras have been aligned in parallel, founding the displacement of the position between the two pixels (one per camera) is reduced to search in each aligned line. That is, if any pair of pixels is visible in both cameras and assuming they are the projection of a single point in the scene, then both pixels must be aligned on the same epipolar line (Ibarra-Manzano, Almanza-Ojeda, Devy, Boizard & Fourniols, 2009). Therefore under this condition, an object location in the scene is reduced to a horizontal translation. Furthermore, the use of the epipolar restriction allows to reduce the complexity and the size of the final architecture.

Next, rectified and corrected images are filtered using an arithmetic mean filter. Once input images have been filtered, they are used to calculate the Census Transform as depicted in the figure 2. This transform is a non-parametric measure used during the matching process for measuring similarities and obtaining the correspondence between the points into the left and right images. A neighborhood of pixels is used for establishing the relationships among them. From the Census Transform, two images are obtained referred to as I_{Cl} and I_{Cr} which represent the left and right Census images. Two pixels extracted from the Census images (one for each image) are compared using the Hamming distance. This comparison

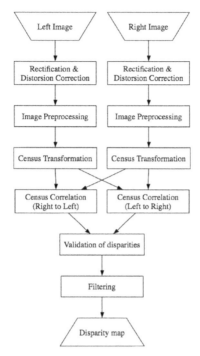

Fig. 2. Block diagram of the stereo vision algorithm (Ibarra-Manzano & Almanza-Ojeda, 2011)

which is called the correlation process allows us to obtain a disparity measure. The disparity measure comes from the similarity maximization function in the same epipolar line for the two Census images. The similarity evaluation is based on the binary comparison between two bit chains calculated by the Census Transform. The correlation process is carried out two times, (left to right then right to left) with the aim of reducing the disparity error and thus complementing the process. Once both disparity measures have been obtained, the disparity measure validation (right to left and left to right) consists of comparing both disparity values and obtaining the absolute difference between them. Before delivering a final disparity image, a novel filtering process is needed for improving its quality. In this final stage, the use of a median spatial filter was found more convenient.

3.1 Disparity map acquired in real time

The final architecture for executing the stereo vision algorithm based on the Census Transform was developed using the level design flow RTL (Ibarra-Manzano, Devy, Boizard, Lacroix & Fourniols, 2009). The architecture was codified in VHDL language using Quartus II workspace and ModelSim. Finally, it was synthesized for an EP2C35F672C6 device contained in the Cyclone IV family of Altera. Table 1 lays out some of the configuration parameters used during the disparity map computation exploited during the dynamic object detection and tracking approach. We would like to highlight that disparity image is computed through the time with a high performance of 30 image per second although detection and tracking approach does not reach this performance. More technical details about the implementation are discussed in section 6.

Parameter	Value
Image size	640×480
Window size	7×7
Disparity max	64
Performance	40
Latency (μs)	206
Area	6,977
Memory size	109 Kb

Table 1. Configuration parameters of the Stereo vision architecture.

The architecture was tested for different navigational scenes using a stereo vision bank, first mounted in a mobile robot and then in a vehicle. In the following, we will describe the obtained results for two operational environments. Figure 3 shows the left and right images acquired from the bank stereo and the associated disparity image delivered by the stereo vision algorithm. Dense disparity image depicts the disparity value in gray color levels in figure 3 (c). By examining this last image, we can determine that if the object is close to the stereo vision bank that means a big disparity value, so it corresponds to a light gray level. Otherwise, if the object is far from the stereo vision bank, the disparity value is low, which corresponds to a dark gray level. In this way, we observe that the gray color which represents the road in the resulting images gradually changes from the light to dark gray level. We point out the right side of the image, where we can see the different tones of gray level corresponding to the building. Since large of the building is located at different depths with respect to the stereo vision bank, in the disparity map corresponding gray color value is assigned from lighter to darker gray tones.

(a) Left image   (b) Right image   (c)   Result   from   FPGA
                                                                    implementation

Fig. 3. Stereo images acquired from a mobile robot during outdoor navigation: a) left image b) right image and c) the disparity map.

In the second test (see figure 4), the stereo vision bank is mounted on a vehicle that is driven on a highway. This experimental test results in a difficult situation because the vehicle is driven at high-speed during the test. Furthermore, this urban condition requires a big distance between both cameras in the stereo vision bank with the aim of augmenting the field of view. A consequence of having a big distance between the two cameras is represented in figure 4 (a) and (b) in which, while the left image show a car that overtakes our vehicle, this car is out of sight in the right camera. Therefore, the dense disparity map shown in figure 4 (c) does not display a cohesive depth value of the vehicle due to mismatches in the information between both images. Nevertheless, we highlight all the different depths represented by the gray color value in the highway that gradually turns darker until the black color which represents an infinity depth.

(a) Left image   (b) Right image   (c)   Result   from   FPGA
                                                                     implementation

Fig. 4. Stereo images acquired from a vehicle in the highway: a) left image b) right image and c) the disparity map.

4. Moving features perception and grouping

4.1 Active 3D point selection
When dealing with vision-based approaches, the problem of processing a large quantity of information requires that the system resources be sophisticated and expensive if we want to get a real time performance. As we work with left and right images from the stereo vision bank, processing both images yields to more information about the scene but has also more computation requirements. To overcome this problem, we consider to deal with a sparse set of features resulted by the analysis of the left image that represents the most significant and salience feature points in all the image. We use this sparse proposition because it is important to distribute time of processing among some others essential tasks of the strategy, therefore feature selection process must expend minimal time. For the feature point selection, we use the Shi-Tomasi approach (Shi & Tomasi, 1994). This approach consists in providing the most representative points based on image gradient analysis, i.e. corners, borders, and all the regions with high contrast change. Figure 5 depicts both image gradients in horizontal and vertical directions of an input image in gray color level, from which N best features are selected for describing image content. In particular, this input image has high content of information, therefore, a larger number of points have to be selected in accordance with image-density information. Furthermore, the number N is restricted by the time of processing required for reaching real-time performance. According to this, the number of points does not have to exceed $N = 180$. Once 2D's interesting points have been selected on the left image and the disparity map for the stereo-images computed, obtaining the 3D characterization of the points is straightforward. Each 2D point is associated with a corresponding depth value provided by the disparity image to conform the 3D point representation.

Until now, we have obtain a set of 2D features on the left image at time t and their corresponding 3D characterization. For each feature, displacement vectors are computed through the time by using the Kanade-Lucas-Tomasi tracker, referred to as the KLT

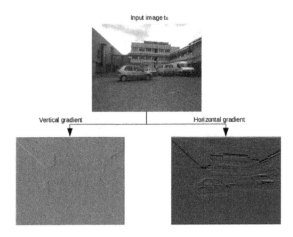

(a) Image gradients

(b) N best features selected

Fig. 5. Image processing for the best interest point selection. a) Gradients in vertical and horizontal direction of the above input image. b) Green points represent the N best interesting features resulted by the gradient image analysis.

technique (Shi & Tomasi, 1994), (Lucas & Kanade, 1981). These displacement vectors are used to calculate feature velocities. We are interested in the accumulation of previous position and velocity of the points in order to establish a trail of motion. An example of the accumulation point positions for N initial features detected appears in figure 6. Note that, while most of the points are distributed in all the image, a small set of points horizontally aligned can be remarked on the left side of the figure c). These points represent the front part of the vehicle that enters into the field of view of our robot. Once a short number of images have been processed, the accumulated vector displacements and positions of the feature points are evaluated in order to find significant patterns of motion that possibly represent dynamic objects in the scene. A new feature selection task is carried out, as indicated in the green block of figure 1. Further, in the case that any dynamic group of points is found, this information will initialize in the second stage of our strategy, that is the object characterization (right rectangular box in figure 1).

In the following, we explain the a *contrario* method used for clustering position and velocity of the points that possibly describe a mobile object.

(a) First image of the tracking   (b) Last image of the tracking   (c)   Accumulate   position
through 4 images

Fig. 6. N feature points initially detected at image a) are tracked through 4 images. Image b) displays the last image during the tracking task. Image c) depicts all the accumulate positions for the N initial points detected calculated by the tracking process.

4.2 Clustering of 3D points cloud

At this point, we have a distributed set of 2D feature points characterized by their position and velocity that have been tracked through a short number of consecutive images. Alternatively, we associated each feature with their corresponding depth position which allows us to manage with a 3D-data set. Following with the global diagram in figure 1, clustering of feature points is the next task to carry out. In order to do that, we use the *a contrario* clustering method proposed in (Veit et al., 2007). This algorithm is based on the Gestalt theory that establishes which groups could be formed based on one or several common characteristics of their elements. In accord to this statement, the *a contrario* clustering technique identifies one group as meaningful if all their elements show a different distribution than an established background random model. Contrary to most clustering techniques, neither initial number of clusters is required nor parameter have to be tuned. These characteristics result very favorable in an unknown environment context where the number of resulted clusters have not been predefined.

In this section we summarize some important concepts of the *a contrario* clustering method, used to group feature points. A detailed description of the method derivation is available in (Desolneux et al., 2008), (Desolneux et al., 2003), (Cao et al., 2007).

4.3 *A contrario* algorithm description

As we mentioned above, a distribution model for the background has to be defined for comparing with the associated distribution of the set of points, here referred to as $V(x,y,v,\theta)$ in R^4. In this work, we use the background model proposed in (Veit et al., 2007) which establishes a random organization of the observations. Therefore, background model elements are independent identically distributed (*iid*) and follow a distribution p. The *iid* nature of random model components proposes an organization with non coherent motion present.

Next, given the input vector $V(x,y,v,\theta)$ from the KLT process (section 4.1), the first objective is to evaluate which elements in V shows a particular distribution contrary to the established distribution p of the background model (that explains "*a contrario*" name). To overcome

(a) Point detection

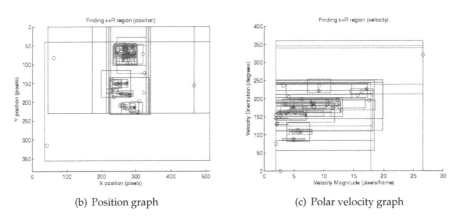

(b) Position graph   (c) Polar velocity graph

Fig. 7. Best fit region search. An initial set of points is selected and tracked in an indoor environment displayed in image a). The accumulated locations and velocities of the points are analyzed in order to find the region of points with $NFA\,(G) \leq 1$.

the problem of point by point evaluation, $V(x,y,v,\theta)$ is divided in testing groups with different size of elements using a single linkage method. This method constructs a binary tree where each node represents a candidate group G. Once a different group of points have been established, these will be evaluate using a set of given regions represented by $\mathcal{H}$. This set of regions is formed by a different size of hyper-rectangles that will be used to test the distribution of each data group in G. An example of groups distribution evaluation is depicted in figure 7. Each region $H \in \mathcal{H}$ is centered at each element $X \in G$ to find the region H_X that contains all the elements in G and at the same time this region has to minimize the probability of the background model distribution. This procedure requires that sizes of hyper-rectangles be in function of data range, in our experiments we use 20 different sizes by dimension. The measure of meaningfulness (called Number of False Alarms NFA in referenced work) is given by eq 1.

$$NFA\,(G) = N^2 \cdot |\mathcal{H}| \min_{\substack{X \in G, \\ H \in \mathcal{H}, \\ G \subset H_X}} B\,(N-1, n-1, p\,(H_X)) \tag{1}$$

In this equation N represents the number of elements in vector V, $|\mathcal{H}|$ is the cardinality of regions and n is the elements in group test G. The term which appears in the minimum function is the accumulated binomial law, this represents the probability that at least n points including X are inside the region test centered in X (H_X). Distribution p consist of four independent distributions, one for each dimension data. Point positions and velocity orientation follow a uniform distribution because object moving position and direction is arbitrary. On the other hand, velocity magnitude distribution is obtained directly of the empirically histogram of the observed data. So that, joint distribution p will be the product of these four distributions. A group G is said to be meaningful if $NFA\,(G) \leq 1$.

Furthermore two sibling meaningful groups in the binary tree could be belong to the same moving object, then a second evaluation for all the meaningful groups is calculated by Eq. 2. To obtain this new measure, we reuse region group information (dimensions and probability) and just a new region that contains both test groups G_1 and G_2 is calculated. New terms are $N' = N - 2$, number of elements in G_1 and G_2, respectively $n'_2 = n_1 - 1$ and $n'_2 = n_2 - 1$, and term $\mathcal{T}$ which represents the accumulated trinomial law.

$$NFA_G\,(G_1, G_2) = N^4 \cdot |\mathcal{H}|^2\,\mathcal{T}\,(N', n'_1, n'_2, p_1, p_2) \tag{2}$$

Both mesures 1 and 2 represent the significance of groups in binary tree. Final clusters are found by exploring all the binary tree and comparing to see if it is more significant to have two moving objects G_1 and G_2 or to fusion it in a group G. Mathematically, $NFA(G) < NFA_G(G_1, G_2)$ where $G_1 \cup G_2 \subset G$. A descriptive result is provided in figure 8. Here blue points correspond to those that could be dynamic but without a well defined motion, so they are associated with the background. On the other hand, green points represent the group in G which shows a different distribution than the random background model. Notice that graph c) displays polar velocity considering magnitude in X-axis and orientation in Y-axis, from there, green point positions are not together because they have orientations among $5°$ and $355°$ Further, these points correspond to the vehicle entering on the right side of the image b).

4.4 Depth validation

As previously mentioned, the 3D characterization of points is achieved using the depth value described in the disparity map for each two dimensional feature point. From previous section, a set of dynamic points have been obtained using the clustering process, however this process was performed considering uniquely 2D points. Thanks to the disparity map, we know the depth of each feature in dynamic group, so it is possible to perform a second evaluation looking for similarity in their depths. This depth evaluation is computed in a formally analogous way to that of the *a contrario* clustering. However in this case, the number of regions for testing is considerably reduced since the group of points is already detected in a defined region, so it is only need the evaluation of similar depths of the points around that region avoiding the best region search. Additionally, there is not an associated velocity in Z-axis direction which reduces from hyper-rectangle regions to 3D boxes. Namely x and y denote the point location in the image and z denotes the disparity value.

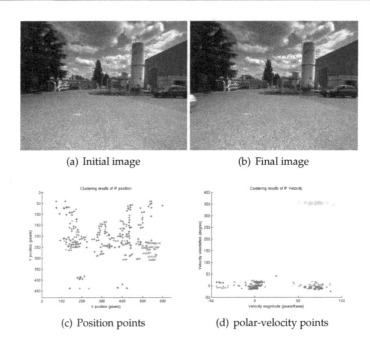

(a) Initial image   (b) Final image

(c) Position points   (d) polar-velocity points

Fig. 8. Clustering results after processing 5 consecutive images. a)First image used for feature selection, b) Last image used for obtaining the trails of the points. A unique dynamic group of points is detected (green points) for whose graph c) depicts their position and graph d) their velocity and orientation.

In the next section, we explain how an irregular contour is initialized using the groups of points calculated by the *a contrario* clustering, therefore used for passing from a dynamic set of points to actual object boundary.

5. Object characterization by an active contour

Active models have been widely used in image processing applications, in particular for recovering shapes and the tracking of moving objects (Li et al., 2010), (Paragios & Deriche, 2005). An active contour or snake is a curve which minimizes energy from restrictive external and internal forces in the image, typically calculated from edges, gradient, among others. Essentially, a snake is not thought of to solve the problem of automatic search for prominent contours of the image, but rather for recovering the contour of a form, from an initial position proposed by other mechanisms. That is to say, if the initial contour is relatively close to the solution (for example a contour defined manually by an operator or obtained through any other method), the contour evolves up until minimizing the function of energy defined from the internal and external forces.

One of the main objectives of this work is to track all mobile objects detected without any prior knowledge about object type that one follows. In a real and dynamic context, we expect to find rigid and non rigid mobile objects ie. vehicles, persons. We have shown in section 4.2 the good performance of the *a contrario* method for finding initial sets of dynamic points which correspond to the mobile objects in the scene. However, whereas clustering process deals with feature points, here we describe a strategy for recovering the deformable shape of the objects through the time by considering others features like intensity image gradient inside of contour. Therefore, through this section we will describe fundamental details of the active contours theory, and the object tracking procedure by means of the active contours.

5.1 Active contour initialization

The results obtained in section 4.2 allow the definition of an initial irregular contour that contains totally or partially our interest object. To this end, we take an outer location of points in the detected group for initializing an active contour which delimits the object on the image. An example is displayed in figure 9 [1]: in image (a) a set of points dynamic (in blue) is detected on which correspond to the person in motion on this sequence. From this set of points we have selected those illustrated in color magenta to describe the object contour depicted in yellow. Due to the fact these points are the most farthest from the center then they are the most representative and closest to the object frontiers. For the results show in image 9(a), we have 8 points on the curve and we use for each 4 control points. The value of 4 control points is fixed for each point on the frontier in order to introduce the corners or discontinuities in the curve (Marin-Hernandez, 2004). Initial contour will allow us to obtain a deformable model by a bounded potential shown in figure 9(b). The zone occupied by the object (represented in red color) is separated from the background. The most intens tones inside of the object are used to best adapt initial contour to the actual object silhouette. Therefore, a more detailed object shape is achieved by analyzing internal and external energy in a bounding box that contains the initial contour.

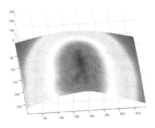

(a) Initial contour of a non-rigid object

(b) energy functional

Fig. 9. Test performed with real images acquired by a fixed camera in an indoor environment. a)Initial contour derived from points further away from the center b) Energy functional that concentrates internal and external energy considering the initial contour of a).

5.1.1 Parametric active contours.

At this point, it is necessary to represent the initial contour in 9 by a parametric curve $u(\tau) = (x(\tau), y(\tau))$, $\tau \in [0, 1]$, with $u(0) = u(1)$. this contour is deformed through the time domain to minimize the energy expressed by:

---

[1] This image sequence was downloaded from the web site (Fisher, 2011) provided by EC Funded CAVIAR project/IST 2001 37540

$$E_{snake} = \int_0^1 \left[E_{int}(u(\tau)) + E_{ext}(u(\tau)) \right] d\tau \tag{3}$$

where E_{int} is expressed by two main terms, the first one refers to the elasticity and the second the flexibility, given:

$$E_{int} = \alpha \int_0^1 |u_\tau(\tau)|^2 d\tau + \beta \int_0^1 |u_{\tau\tau}(\tau)|^2 d\tau \tag{4}$$

τ and $\tau\tau$ indexes in the term $u(\tau)$ implies respectively first and second order of derivation. By returning to the equation 3 for defining the term E_{ext} or the energy of the image (Sekhar et al., 2008), as the field of potential P:

$$E_{ext} = \int_0^1 P(u(\tau)) d\tau \tag{5}$$

The potential includes different terms defined from image proprieties like edges, lines, etc. Edges energy is obtained by computing the magnitude of the gradient intensity $|\nabla I|$. Without a good initialization, the energy of the edges will not be enough to locate the objects on noisy or low contrast images. Therefore an additional potential of the regions is added to the edge energy. Generally, the potential of the regions is defined by the mean (μ) and the variance (σ^2) of the pixels intensity in the region. However, other constraints could be added like the object velocities, or other statistics derived from region characteristics (Brox et al., 2010). Because of the real-time constraints, we calculate only some statistics in the region that describe the object, such as the main properties to its implementation in correspondence to the next image. The following section describes the proposed strategy about shape recovering and object tracking using active contours.

5.2 Incremental silhouette definition

We have tested a method based on the work of Chan and Vese (Chan & Vese, 2001) in order to find the silhouette of the object. In this work, the authors give a region of initialization which may contain total or partially the object. The analysis of this initial contour will allow evaluation of the conditions of the energy minimization inside and outside of the contour. An example of a shape recovering of a person in an indoor environment using the Chan and Vese method is shown in figure 10. First row of images display the contour evolution. In the second row the region inside of the contour for each corresponding above image is illustrated. These images are given as input to the process for recovering real object contour. Furthermore, white regions on these images are labeled as occupied locations which avoid the detection of new interesting points inside of it through the time. According to this, new feature selection (detailed in section 4.1) will look for unoccupied locations in the image allowing the detection of incoming objects.

Once a partial contour at image t in figure 10 has been obtained by minimizing eq. 3, we estimate its position at image $t + 1$ for starting a new convergence process. The prediction of the region on the next image is always given by the Kalman filter. The vector state of our object is expressed as:

$$\mathbf{x}_0 = \left[\bar{x}, \bar{y}, \bar{v}_x, \bar{v}_y \right]^T \tag{6}$$

Namely, $\bar{x}$, $\bar{y}$ denote the barycenter location and $\bar{v}_x$, $\bar{v}_y$ the means of velocity vector in X and Y direction respectively. Figure 11 illustrates an example of the filter prediction. The barycenter position and the partial converged contour are located in accordance with Kalman

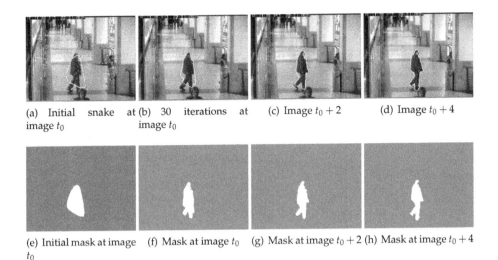

(a) Initial snake at image t_0 (b) 30 iterations at image t_0 (c) Image $t_0 + 2$ (d) Image $t_0 + 4$

(e) Initial mask at image t_0 (f) Mask at image t_0 (g) Mask at image $t_0 + 2$ (h) Mask at image $t_0 + 4$

Fig. 10. Results of the detection and tracking of a non-rigid dynamic object, in this case a pedestrian on the same sequence of figure 9. Here, we have processed 5 consecutive images, executing 30 iterations per image in order to find the actual object shape.

filter prediction. We always consider a velocity constant model and the vector state is obtained from model object displacement. However, in some cases it is first necessary to tune the initial parameters of the filter because there may exist different types of object movements producing an undesired acceleration of the predicted position.

5.3 Object contour validation

The difficulties rising from our proposed strategy points out the sensitivity of the active contours to small discontinuities on the object edges with low-contrast because the energy is configured over the entire image. To overcome these difficulties we evaluate the disparity inside of the region corresponding with the binary mask (that will be presented in the next section figure 15). This evaluation consists in ordering in an ascendant way all the disparity values inside the region designed by the binary mask then we uniquely consider the median of the values. In particular, we consider a valid point of the object region if its disparity value is located up to the 4th percentil in the ordered list (here referred to as $4p$). It follows that our statistical validation for object contour refining can be written as:

$$ M = \begin{cases} 1 & depth \geq 4p \\ 0 & otherwise \end{cases} \tag{7} $$

where M represents the binary mask. It is important to remark that the depth value is the inverse value of the disparity therefore in this case the fact of rejecting points located on the first four percentiles represents that these values have a lower disparity value, that is, disparity values of the background points are expecting to be low. This constrain allow us to develop a last validity evaluation for obtaining a most detailed and accurated representation of the object shape. The next section describes our experimental results and presents how the disparity map plays a fundamental role in increasing the efficiency and improve the obtained results.

(a) Object silhouette   (b) Estimation for the next image

Fig. 11. Motion estimation using Kalman filter. a) The barycenter location at image t is used for predict its position at the next image. b) The irregular contour is centered at the predicted barycenter position, this will be used as initialization region in incoming images.

6. Experimental results

We evaluate the performance of the proposed strategy for detecting and tracking dynamic objects by carrying out experiments using both secure and mounted cameras on autonomous vehicles. Since, using one secure camera, vibrations, egomotion, among other typical problems of mounted cameras are neglected, first we do an experiment under this controlled condition to verify that our algorithm detects moving objects in real images. Moreover, we have observed that the total convergence of energy equation is computationally demanding. To overcome this constraint, it was necessary to propose that convergence task works uniquely a small number of iterations and re-starting the process at next image from previous results. In our case, we see a significant improvement in the efficiency of computation by using this, even when the most similar contour to actual object shape is found after some iterations.

6.1 Fixed camera case
Figure 12 shows 10 consecutive images in which a person appears in the field of view of one security camera on a Commercial Center. These resulting images refer to the case where the number of iterations for converging the active contour at each image is set to 30. The initial contour obtained from the cluster of dynamic points was displayed in figure 9. The effect of the Kalman filter estimation permit us to achieve a more detailed shape than the initial contour frame by frame. We remark that even if initial contour does not contain all the object, the actual shape is achieved after processing 5 images. It is important to point out that resulting object bounds are almost the same as the real thanks to the high-contrast generated between the person and the background.

6.2 Experiments during mobile robot navigation: rigid objects detection
A second experiment was performed in an outdoor environment where the robot accomplishes a linear trajectory of navigation at low-speed (about 6 m/s). In our experiments, we only use the stereo-vision bank mounted on the mobile robot. The robot cameras provide images with resolution of 640×480 pixels. The egomotion derived from the robot displacements is neglected by verifying similar depth values inside of the region containing the dynamic group of points. Moreover, dynamic groups detected on the road shows different depth values, so they are rejected to define an initial contour. Figure 13 illustrates the detection and tracking of a white vehicle. Note that this vehicle comes from the imaginary epipolar

Fig. 12. Detection and tracking of a non-rigid dynamic object along 12 consecutive images. In each image, 30 iterations are used in order to find the object silhouette.

point of image, however it can only be perceived by our strategy at the moment of it is closer to our robot position. Whereas the first experiment was performed under ideal conditions of controlled illumination provided by the indoor environment here this conditions do not hold. This fact notably avoids the computation of a cohesive representation of the energy functional as the one illustrated in figure 9(b). As a consequence, there is only one object at each image of figure 13 but its contour is represented by two separated regions that have similar energy level inside. Furthermore, we analyze the disparity map for obtaining the statistics of disparity values in both regions as mentioned in section 5.3

(a)   (b)   (c)   (d)

(e)   (f)   (g)   (h)

Fig. 13. Experimental results in an outdoor environment during robot navigation. Images show the detection and tracking of a rigid object.

6.3 Experiments during mobile robot navigation: non rigid objects detection

In the third experiment, we considered a mobile robot moving in an outdoor environment again but in this case the robot finds a dynamic non-rigid object during its trajectory. The left image of figure 14 displays the frame in which the dynamic object was detected and initialized as a irregular contour, middle image shows the disparity map and right image the respective initial mask. In practice detecting non-rigid objects is more complicated than the previous experiment because a person walking has different motions in his legs than his shoulder or his head. Because of improvements, we use in this experiment 40 iterations per image for converging the active contour. Figure 15 illustrates some resulted images of the tracking performed by our proposed strategy. In this experiment, we found that almost all the person could be covered with the active contour. However by examining the left column

(a) Initial detection   (b) disparity map   (c) Initial mask

Fig. 14. a) Initial contour derived from the dynamic group of points. b) The corresponding disparity map of the image, c) the initial mask used by the active contour.

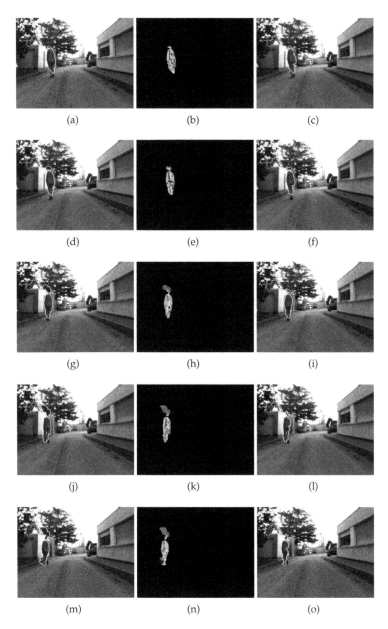

(a)   (b)   (c)

(d)   (e)   (f)

(g)   (h)   (i)

(j)   (k)   (l)

(m)   (n)   (o)

Fig. 15. Images on the left column show the contour obtained by only considering the intensity inside of the region. The middle column isolates the object contour representation and its corresponding disparity values. Images on the right evaluate the depth of the target to boil down extra regions included in the object contour.

of figure 15, we can figure out that part of the background was also included in our object region. To overcome this problem we performed the same statistical calculation of disparity values inside of the region-object as described in section 5.3. By examining the middle column of figure 15, we can figure out that similar depth values are concentrated in actual object contour. We can see that additional zones to this contour that represents the tree and other locations of the background will be rejected using the disparity map since they provide lower values of disparity, that is they are farther away than our interesting moving object.

The proposed algorithm was coded on C/C++ and TCL, however disparity maps are computed for the FPGA by means of an architecture codified in VHDL. The Cyclone IV card calculates 30 disparity maps per second. After several tests, we measure that our algorithm runs around 1 or 1.5 Hz depending of the nature of the environment. From this, note that the disparity is available at higher frequency than our algorithm performance, however we comment that until now the goal of the experiment was to provide an algorithm for detecting and tracking moving objects.

7. Conclusions

In this project, we considered the problem of dynamic object detection from a mobile robot in indoor/outdoor environments of navigation. Specially, the proposed strategy uses only visual-information provided by a stereo-vision bank mounted on the mobile robot. The speed of the robot during navigation is established low to avoid disturbance on the velocity data due to robot ego-motion. Experimental results allow us to realize that the proposed strategy performs a correct and total detection of the rigid and non-rigid objects and it is able to tracking them along the image sequence. Motivated by these results future contributions to this project consist in decreasing the time of computation. Nevertheless, we make some assumptions by avoiding excessive or unnecessary computations ie. the number of selected feature points, number of iterations during the active contour processing, our global algorithm is not able to perform in real time (at least 10 Hz). Significant improvements could be obtained by an emigration of all our algorithm design to embedded architectures like GPU or FPGA devices. Furthermore these kinds of devices provide a high portability towards robotics or autonomous vehicle platforms.

We also comment the difficulties rising from the disparity map constructed by the stereo vision module in which a cohesive and accurate representation of the actual scene have to be improved. To this end, future works consider the addition of a strategy for rejecting "spikes" in the disparity map cased by stereo mismatches.

8. Acknowledgments

This work was partially funded by the CONACyT with the project entitled "Diseno y optimizacion de una arquitectura para la clasificacion de objetos en tiempo real por color y textura basada en FPGA". Authors would like to thank Cyril Roussillon for help provided with the robot.

9. References

Almanza-Ojeda, D., Devy, M. & Herbulot, A. (2010). Visual-based detection and tracking of dynamic obstacles from a mobile robot, *In Proceedings of the 7th International Conference on Informatics in Control, Automation and Robotics (ICINCO 2010)*, Madeira, Portugal.

Almanza-Ojeda, D. L., Devy, M. & Herbulot, A. (2011). Active method for mobile object detection from an embedded camera, based on a contrario clustering, *in* J. A. Cetto, J.-L. Ferrier & J. Filipe (eds), *Informatics in Control, Automation and Robotics*, Vol. 89 of *Lecture Notes in Electrical Engineering*, Springer Berlin Heidelberg, pp. 267–280. 10.1007/978-3-642-19539-6_18.
URL: *http://dx.doi.org/10.1007/978-3-642-19539-6_18*

Brox, T., Rousson, M., Deriche, R. & Weickert, J. (2010). Colour, texture, and motion in level set based segmentation and tracking, *Image Vision Comput.* 28(3): 376–390.

Cao, F., Delon, J., Desolneux, A., Musé, P. & Sur, F. (2007). A unified framework for detecting groups and application to shape recognition, *Journal of Mathematical Imaging and Vision* 27(2): 91–119.

Chan, T. & Vese, L. (2001). Active contours without edges, *Transactions on Image Processing, IEEE* 10(2): 266–277.

Desolneux, A., Moisan, L. & Morel, J.-M. (2003). A grouping principle and four applications, *IEEE Transactions on Pattern Analysis and Machine Intelligence* 25(4): 508–513.

Desolneux, A., Moisan, L. & Morel, J.-M. (2008). *From Gestalt Theory to Image Analysis A Probabilistic Approach*, Vol. 34, Springer Berlin / Heidelberg.

Fisher, R. (2011).
URL: *http://groups.inf.ed.ac.uk/vision/CAVIAR/CAVIARDATA1/*

Ibarra-Manzano, M.-A. & Almanza-Ojeda, D.-L. (2011). *Advances in Stereo Vision*, InTech, chapter High-Speed Architecture Based on FPGA for a Stereo-Vision Algorithm, pp. 71-88.

Ibarra-Manzano, M., Almanza-Ojeda, D.-L., Devy, M., Boizard, J.-L. & Fourniols, J.-Y. (2009). Stereo vision algorithm implementation in fpga using census transform for effective resource optimization, *Digital System Design, Architectures, Methods and Tools, 2009. 12th Euromicro Conference on*, pp. 799 –805.

Ibarra-Manzano, M., Devy, M., Boizard, J.-L., Lacroix, P. & Fourniols, J.-Y. (2009). An efficient reconfigurable architecture to implement dense stereo vision algorithm using high-level synthesis, *2009 International Conference on Field Programmable Logic and Applications*, Prague, Czech Republic, pp. 444–447.

Katz, R., Douillard, B., Nieto, J. & Nebot, E. (2008). A self-supervised architecture for moving obstacles classification, *IEEE/RSJ International Conference on Intelligent Robots and Systems, IROS 2008*, pp. 155–160.

Klappstein, J., Vaudrey, T., Rabe, C., Wedel, A. & Klette, R. (2008). Moving object segmentation using optical flow and depth information, *PSIVT '09: Proceedings of the 3rd Pacific Rim Symposium on Advances in Image and Video Technology*, Springer-Verlag, Berlin, Heidelberg, pp. 611–623.

Li, C., Xu, C., Gui, C. & Fox, M. D. (2010). Distance regularized level set evolution and its application to image segmentation, *IEEE Trans. Image Process.* 19(12): 3243–3254.

Lookingbill, A., Lieb, D. & Thrun, S. (2007). *Autonomous Navigation in Dynamic Environments*, Vol. 35 of *Springer Tracts in Advanced Robotics*, Springer Berlin / Heidelberg, pp. 29–44.

Lucas, B. D. & Kanade, T. (1981). An iterative image registration technique with an application to stereo vision, *Proceedings of DARPA Image Understanding Workshop*, pp. 121–130.

Marin-Hernandez, A. (2004). *Vision dynamique pour la navigation d'un robot mobile.*, PhD thesis, INPT-LAAS-CNRS.

Masrani, D. & MacLean, W. (2006). A Real-Time large disparity range Stereo-System using FPGAs, *Computer Vision Systems, 2006 ICVS '06. IEEE International Conference on*, p. 13.

Matthies, L., Litwin, T., Owens, K., Rankin, A., Murphy, K., Coombs, D., Gilsinn, J., Hong, T., Legowik, S., Nashman, M. & Yoshimi, B. (1998). Performance evaluation of ugv obstacle detection with ccd/flir stereo vision and ladar, *ISIC/CIRA/ISAS Joint Conference* pp. 658–670.

Paragios, N. & Deriche, R. (2005). Geodesic active regions and level set methods for motion estimation and tracking, *Computer Vision and Image Understanding* 97(3): 259 – 282. URL: *http://www.sciencedirect.com/science/article/pii/S1077314204001213*

Schmit, H. H., Cadambi, S., Moe, M. & Goldstein, S. C. (2000). Pipeline reconfigurable fpgas, *Journal of VLSI Signal Processing Systems* 24(2-3): 129–146.

Sekhar, S. C., Aguet, F., Romain, S., Thévenaz, P. & Unser, M. (2008). Parametric b-spline snakes on distance maps—application to segmentation of histology images, *Proceedings of the 16th European Signal Processing Conference, (EUSIPCO2008)* .

Shi, J. & Tomasi, C. (1994). Good features to track, *proceedings of the IEEE Conference on Computer Vision and Pattern Recognition*, pp. 593–600.

Sola, J., Monin, A. & Devy, M. (2007). BiCamSLAM: two times mono is more than stereo, *IEEE International Conference on Robotics Automation (ICRA2007), Rome, Italy*, pp. 4795–4800.

Talukder, A. & Matthies, L. (2004). Real-time detection of moving objects from moving vehicles using dense stereo and optical flow, *proceedings of the International Conference on Intelligent Robots and Systems (IROS2004)*, pp. 3718–3725.

Veit, T., Cao, F. & Bouthemy, P. (2007). Space-time a contrario clustering for detecting coherent motion, *IEEE International Conference on Robotics and Automation, (ICRA07)*, Roma, Italy, pp. 33–39.

Vu, T. & Aycard, O. (2009). Laser-based detection and tracking moving objects using data-driven markov chain monte carlo, *IEEE International Conference on Robotics Automation (ICRA2009), Kobe, Japan*.

Williamson, T. (1998). *A High-Performance Stereo Vision System for Obstacle Detection*, PhD thesis, Robotics Institute, Carnegie Mellon University, Pittsburgh, PA.

Woodfill, J., Gordon, G., Jurasek, D., Brown, T. & Buck, R. (2006). The tyzx DeepSea g2 vision system, ATaskable, embedded stereo camera, *Computer Vision and Pattern Recognition Workshop, 2006. CVPRW '06. Conference on*, p. 126.

Zabih, R. & Woodfill, J. (1994). Non-parametric local transforms for computing visual correspondence, *ECCV '94: Proceedings of the Third European Conference on Computer Vision*, Vol. II, Springer-Verlag New York, Inc., Secaucus, NJ, USA, pp. 151–158.

4

Three-Dimensional Environment Modeling Based on Structure from Motion with Point and Line Features by Using Omnidirectional Camera

Ryosuke Kawanishi, Atsushi Yamashita and Toru Kaneko

Shizuoka University

Japan

1. Introduction

Three-dimensional map is available for autonomous robot navigation (path planning, self-localization and object recognition). In unknown environment, robots should measure environments and construct their maps by themselves.

Three-dimensional measurement using image data makes it possible to construct an environment map (Davison, 2003). However, many environmental images are needed if we use a conventional camera having a limited field of view (Ishiguro et al., 1992). Then, an omnidirectional camera is available for wide-ranging measurement, because it has a panoramic field of view (Fig. 1). Many researchers showed that an omnidirectional camera is effective in measurement and recognition in environment (Bunschoten & Krose, 2003; Geyer & Daniilidis, 2003; Gluckman & Nayar, 1998).

Hyperboloid mirror

Camera

Fig. 1. Omnidirectional camera equipped with a hyperboloid mirror. The left figure shows an acquired image.

Our proposed method is based on structure from motion. Previous methods based on structure from motion often use feature points to estimate camera movement and measure environment (Rachmielowski et al., 2006; Kawanishi et al., 2009). However, many non-textured objects may exist in surrounding environments of mobile robots. It is hard to extract enough number of feature points from non-textured objects. Therefore, in an environment having non-textured objects, it is difficult to construct its map by using feature points only.

Then, line features should be utilized for environment measurement, because non-texture objects often have straight-lines. As examples of previous works using lines, a method for precise camera movement estimation by using stereo camera (Chandraker et al., 2009), a method for buildings reconstruction by using orthogonal lines (Schindler, 2006) and so on (Bartoli & Sturm, 2005; Smith et al., 2006; Mariottini & Prattichizzo, 2007) have been proposed.

However, there is a prerequisite on previous line detections of them. A method must obtain a vanishing point (Schindler, 2006) or a pair of end points of the straight-line (Smith et al., 2006). Some of previous line detection is only for a normal camera (Chandraker et al., 2009). Alternatively, some previous methods obtain line correspondences by hand (Bartoli & Sturm, 2005; Mariottini & Prattichizzo, 2007).

We propose a method for straight-line extraction and tracking on distorted omnidirectional images. The method does not require a vanishing point and end points of straight-lines. These straight-lines are regarded as infinite lines in the measurement process (Spacek, 1986). Therefore, the proposed method can measure straight-lines even if a part of the line is covered by obstacles during its tracking.

Our proposed method measures feature points together with straight-lines. If only straight-lines are used for camera movement estimation, a non-linear problem must be solved. However, camera movement can be estimated easily by a linear solution with point correspondences. Moreover, although few numbers of straight-lines may be extracted from textured objects, many feature points will be extracted from them. Therefore, we can measure the environment densely by using both feature points and straight-lines.

The process of our proposed method is mentioned below (Fig. 2). First, feature points and straight-lines are extracted and tracked along an acquired omnidirectional image sequence. Camera movement is estimated by point-based Structure from Motion. The estimated camera movement is used for an initial value for line-based measurement.

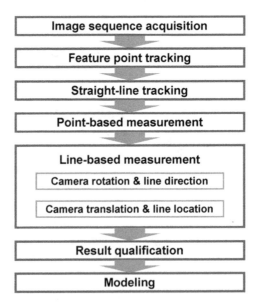

Fig. 2. Procedure of our proposed method.

The proposed line-based measurement is divided into two phases. At the first phase, camera rotation and line directions are optimized. Line correspondence makes it possible to estimate camera rotation independently of camera translation (Spacek, 1986). Camera rotation can be estimated by using 3-D line directions. At the second phase, camera translation and 3-D line location are optimized. The optimization is based on Bundle adjustment (Triggs et al., 1999). Some of measurement results have low accuracy. The proposed method rejects such results.

Measurement results of feature points and straight-lines are integrated. Triangular meshes are generated from the integrated measurement data. By texture-mapping to these meshes, a three-dimensional environment model is constructed.

2. Coordinate system of omnidirectional camera

The coordinate system of the omnidirectional camera is shown in Fig. 3. A ray heading to image coordinates (u, v) from the camera lens is reflected on a hyperboloid mirror. In this paper, the reflected vector is called a ray vector. The extension lines of all ray vectors intersect at the focal point of the hyperboloid mirror. The ray vector $\mathbf{r}$ is calculated by the following equations.

$$\mathbf{r} = \begin{bmatrix} \lambda(u - c_x)p_x \\ \lambda(v - c_y)p_y \\ \lambda f - 2\gamma \end{bmatrix} \tag{1}$$

$$\lambda = \frac{\alpha^2 \left(f\gamma + \beta\sqrt{u^2 + v^2 + f^2} \right)}{\alpha^2 f^2 - \beta^2 \left(u^2 + v^2 \right)} \tag{2}$$

where c_x and c_y are the center coordinates of the omnidirectional image, p_x and p_y are pixel size, f is the focal length, α, β and γ are hyperboloid parameters. These parameters are calibrated in advance.

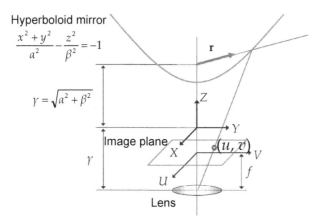

Fig. 3. The coordinate system of the omnidirectional camera. Ray vector $\mathbf{r}$ is defined as a unit vector which starts from the focal point of a hyperboloid mirror.

3. Feature tracking

3.1 Point tracking

Feature points are tracked along an omnidirectional image sequence by KLT tracker (Shi & Tomasi, 1994). These points are used for initial estimation of camera movement and measurement for textured objects. An example of feature point tracking is shown in Fig. 4.

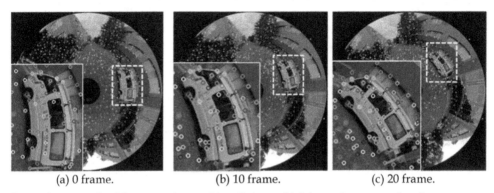

(a) 0 frame.   (b) 10 frame.   (c) 20 frame.

Fig. 4. An example of feature point tracking. Points which have the same color show corresponding points.

3.2 Straight-line tracking

Straight-lines are extracted from a distorted omnidirectional image. The proposed method obtains edge points by Canny edge detector (Canny, 1986). An example of edge point detection is shown in Fig. 5 (a) and (b).

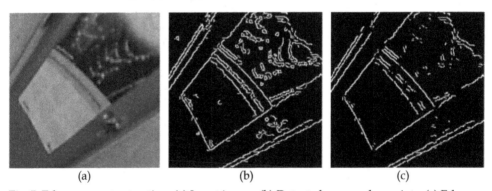

(a)   (b)   (c)

Fig. 5. Edge segment extraction. (a) Input image. (b) Detected canny edge points. (c) Edge segments are separated by rejecting corner points.

To separate each straight-line, corner points are rejected as shown in Fig. 5 (c). Corner points are detected by using two eigenvalues of the Hessian of the image. Hessian matrix is calculated by the following equation.

$$\mathbf{H} = \begin{bmatrix} I_x^2 & I_x I_y \\ I_x I_y & I_y^2 \end{bmatrix} \tag{3}$$

Three-Dimensional Environment Modeling Based on Structure from Motion with Point and
Line Features by Using Omnidirectional Camera

69

where I_x and I_y are derivatives of image I. An edge point which has large value of the ratio of eigenvalues is regarded as a point locating on a line. In the proposed method, if the ratio is smaller than 10, the edge point is rejected as a corner point. This process provides us separated edge segments.

A least square plane is calculated from ray vectors of edge points which belong to an edge segment. If the edge segment constitutes a straight-line, these ray vectors are located on a plane (Fig. 6). Therefore, an edge segment which has a small least square error is regarded as a straight-line. The proposed method is able to extract straight-lines, even if an edge segment looks like a curved line in a distorted omnidirectional image.

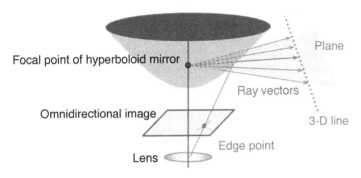

Fig. 6. The relationship between a straight-line and a ray vector.

The maximum number of edge points which satisfy equation (4) is calculated by using RANSAC (Fischler & Bolles, 1981).

$$\left(\mathbf{r}_{i,j}{}^{T}\mathbf{n}_{i}\right)^{2} < l_{\text{th}} \tag{4}$$

where l_{th} is a threshold. $\mathbf{r}_{i,j}$ is a ray vector heading to an edge point j included in an edge segment i. $\mathbf{n}_i$ is the normal vector of the least square plane calculated from the edge segment i. If over half the edge points of the edge segment i satisfy equation (4), the edge segment is determined as a straight-line. The threshold l_{th} is calculated by the following equation.

$$l_{\text{th}} = \cos^{2}\left(\frac{2\pi}{2r_{\text{m}}\pi}\right) = \cos^{2}\left(\frac{1}{r_{\text{m}}}\right) \tag{5}$$

where r_{m} is the radius of projected mirror circumference in an omnidirectional image. A threshold l_{th} allows angle error within $1/r_{\text{m}}$ [rad]. It means that an angle error between a ray vector and a straight-line is within 1 pixel.

Straight-lines are tracked along an omnidirectional image sequence. The proposed method extracts points at constant intervals on a straight-line detected in the current frame (Fig. 7 (a) and (b)). These points are tracked to the next frame by KLT tracker (Fig. 7 (d)). Edge segments are detected in the next frame (Fig. 7 (c)). The edge point closest to the tracked point is selected as a corresponding edge point (Fig. 7 (e)). The edge segment which has themaximum number of corresponding edge points is regarded as a corresponding edge segment (Fig. 7 (f)). If an edge segment corresponds to several lines, a line which has larger number of corresponding edge points is selected.

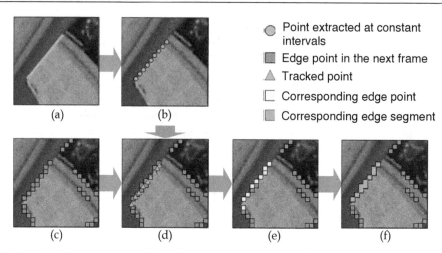

Point extracted at constant intervals

Edge point in the next frame

Tracked point

Corresponding edge point

Corresponding edge segment

(a)   (b)

(c)   (d)   (e)   (f)

Fig. 7. Searching for a corresponding edge segment in the next frame. (a) Straight-line extracted in the current frame. (b) Points extracted at constant intervals on the line. (b) Edge segments in the next frame. (c) Points (b) are tracked between the current frame and the next frame. (d) Corresponding edge points. (e) Corresponding edge segment.

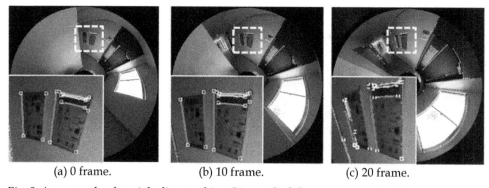

(a) 0 frame.   (b) 10 frame.   (c) 20 frame.

Fig. 8. An example of straight-line tracking. Lines which have a same color show corresponding lines. Although an end point of a line is shown as a white square, it is not used for straight-line detection.

Matching point search on a line has the aperture problem (Nakayama, 1988). However, it is not difficult for the proposed method to obtain corresponding lines, because it does not require point-to-point matching. By continuing the above processes, straight-lines are tracked along the omnidirectional image sequence. An example of line tracking is shown in Fig. 8.

4. Environment measurement

4.1 Point-based measurement

Camera movement is estimated by a point-based method (Kawanishi et al., 2009). The method is based on eight-point algorithm (Hartley, 1997).

An essential matrix $\mathbf{E}$ is calculated from ray vectors of corresponding feature points. An essential matrix $\mathbf{E}$ and ray vectors satisfy the following equation.

$$\mathbf{r}_i \mathbf{E} \mathbf{r}'_i = 0 \tag{6}$$

where ray vectors $\mathbf{r}_i = \left[x_i, y_i, z_i \right]^{\mathrm{T}}$ and $\mathbf{r}'_i = \left[x'_i, y'_i, z'_i \right]^{\mathrm{T}}$ are those of the corresponding point in two images. Camera rotation matrix $\mathbf{R}$ and translation vector $\mathbf{t}$ are calculated from essential matrix $\mathbf{E}$ by singular value decomposition. Equation (6) is transformed as follows.

$$\mathbf{u}^{\mathrm{T}} \mathbf{e} = 0 \tag{7}$$

where

$$\mathbf{u} = \left[x_i x'_i, y_i x'_i, z_i x'_i, x_i y'_i, y_i y'_i, z_i y'_i, x_i z'_i, y_i z'_i, z_i z'_i \right]^{\mathrm{T}},$$

$$\mathbf{e} = \left[e_{11}, e_{12}, e_{13}, e_{21}, e_{22}, e_{23}, e_{31}, e_{32}, e_{33} \right]^{\mathrm{T}}.$$

e_{ab} is the row a and column b element of Essential matrix $\mathbf{E}$. The matrix $\mathbf{E}$ is obtained by solving simultaneous equations for more than eight pairs of corresponding ray vectors.

$$J = \left\| \mathbf{U} \mathbf{e} \right\|^2 \rightarrow \min \tag{8}$$

where $\mathbf{U} = \left[\mathbf{u}_1, \mathbf{u}_2, \cdots, \mathbf{u}_n \right]^{\mathrm{T}}$. n is the number of feature points. $\mathbf{e}$ is calculated as the eigenvector of the smallest eigenvalues of $\mathbf{U}^{\mathrm{T}} \mathbf{U}$. Estimated camera movement in this process is used as an initial value for line-based measurement. However, not all feature points tracked in the image sequence correspond satisfactorily due to image noise, etc. Mistracked feature points should be rejected. The proposed method rejects these points as outliers by using RANSAC algorithm (Fischler & Bolles, 1981).

4.2 Line-based measurement
Estimated camera movement is optimized by using straight-lines. A straight-line is represented as infinite lines by using its direction vector $\mathbf{d}^w$ and location vector $\mathbf{l}^w$ ($\mathbf{l}^w + k\mathbf{d}^w$, k is a factor). The superscript w means that the vector is in world coordinate system. As a prerequisite for line-based measurement, at least, more than 3 images and 3 pairs of corresponding lines (at least one line is not parallel to others) are needed. In the first step, camera rotation and line directions are estimated. The step is independent of camera translation and line locations estimation. In the next step, camera translation and line locations are optimized by a method based on Bundle adjustment (Triggs et al., 1999). In these phases, initial value of 3-D line direction and location are required. These initial values are calculated from line correspondences and initial camera movements.

4.2.1 Camera rotation and 3-D line direction optimization
Our proposed method calculates a normal vector $\mathbf{n}_i^c$ of a least square plane calculated from an edge segment i in Section 3.2. The superscript c means that the vector is in a camera coordinate system at camera position c. Camera rotation depends on 3-D line direction vector $\mathbf{d}_i^w$ and normal vector $\mathbf{n}_i^c$. By using initial values of camera rotation and normal

vectors $\mathbf{n}_i^c$, a sum of errors E_R between camera rotation matrix $\mathbf{R}_c^w$ and 3-D line direction $\mathbf{d}_i^w$ are calculated as shown in the following equation.

$$E_R = \sum_c \sum_i \left(\left(\mathbf{R}_c^{wT} \mathbf{n}_i^c \right)^T \mathbf{d}_i^w \right)^2 \qquad (9)$$

where, $\mathbf{R}_c^w$ is a rotation matrix from the world coordinate system to camera coordinate system c. Here, $\mathbf{d}_i^w$ and $\mathbf{n}_i^c$ are unit vectors. The relationship between a direction vector and a normal vector is shown in Fig. 9. Camera rotations and line directions are optimized by minimizing E_R. Levenburg-Marquardt method is used for the minimization.

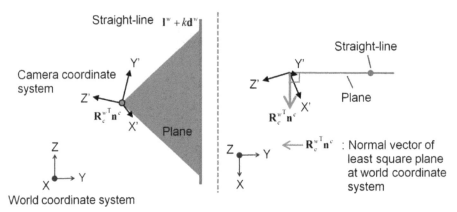

Fig. 9. Relationship between a direction vector of straight-line and a normal vector of a least square plane.

4.2.2 Camera translation and 3-D line location optimization

Camera translation vector $\mathbf{t}_c^w$ and 3-D line location $\mathbf{l}_i^w$ are optimized by Bundle adjustment (Triggs et al., 1999). The method estimates camera movements by minimizing reprojection errors. The projection error of the straight-line is calculated as an angle error between two vectors on a plane which is orthogonal to the line direction. The sum of reprojection errors of straight-lines E_t is calculated by the following equation.

$$E_t = \sum_c \sum_i \left(1 - \hat{\mathbf{l}}_i^{cT} \mathbf{g}_i^c \right)^2 \qquad (10)$$

where $\mathbf{g}_i^c$ is a vector located on the plane $\mathbf{n}_i^c$, and it crosses the 3-D line at a right angle. Thus, $\mathbf{g}_i^c$ satisfies $\mathbf{R}_c^{wT} \mathbf{g}_i^c \perp \mathbf{d}_i^w$ and $\mathbf{g}_i^c \perp \mathbf{n}_i^c$. $\hat{\mathbf{l}}_i^c$ is a vector which connects the camera position c and the 3-D line location with the shortest distance. $\hat{\mathbf{l}}_i^c$ is calculated by the following equation.

$$\hat{\mathbf{l}}_i^c = \mathbf{R}_c^w \left(B_{i,c} \mathbf{d}_i^w + \mathbf{l}_i^w - \mathbf{t}_c^w \right) \Big/ \left\| B_{i,c} \mathbf{d}_i^w + \mathbf{l}_i^w - \mathbf{t}_c^w \right\| \qquad (11)$$

where $B_{i,c}$ is a factor which shows a location on the 3-D line. $B_{i,c}$ satisfies the following equation.

Three-Dimensional Environment Modeling Based on Structure from Motion with Point and
Line Features by Using Omnidirectional Camera

73

$$\left\|\left(A_{i,c}\mathbf{R}_c^{w\mathrm{T}}\mathbf{g}_i^c + \mathbf{t}_c^w\right) - \left(B_{i,c}\mathbf{d}_i^w + \mathbf{1}_i^w\right)\right\| \to \min \tag{12}$$

The relationship between these vectors is shown in Fig. 10.

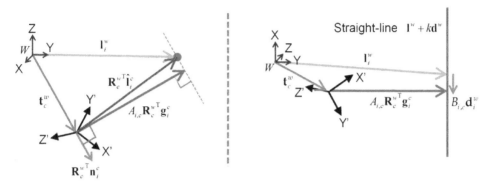

Fig. 10. Relationship between camera translation vector and 3-D line location vector.

The sum of reprojection errors of straight-lines E_t is minimized by a convergent calculation based on Levenburg-Marquardt method. In these two optimization steps, lines which have large error are rejected as outliers by RANSAC algorithm.

In the proposed method, 3-D lines are represented as uniformly-spaced points $\mathbf{e}_{i,n}^w$. 3-D coordinates of these points are calculated by the following equation.

$$\mathbf{e}_{i,n}^w = hn\mathbf{d}_i^w + \mathbf{1}_i^w \tag{13}$$

where h is a uniform distance and n is an integer number, respectively. 3-D coordinates of $\mathbf{e}_{i,n}^w$ is reprojected to the image sequence. When the 2-D coordinates of the reprojection point are close to the corresponding edge segment enough, the point is added into measurement data.

By using estimated camera movement, 3-D coordinates of feature points which have the minimal reprojection error are calculated and integrated with straight-line measurement data.

4.3 Result qualification
Measurement data which have low accuracy should be rejected before 3-D model construction. Measurement accuracy of the feature point is evaluated by following equations.

$$a_{i,m} = \left\|\frac{\partial \mathbf{p}_{i,m}}{\partial u_{c1,i}}\right\| + \left\|\frac{\partial \mathbf{p}_{i,m}}{\partial v_{c1,i}}\right\| + \left\|\frac{\partial \mathbf{p}_{i,m}}{\partial u_{c2,i}}\right\| + \left\|\frac{\partial \mathbf{p}_{i,m}}{\partial v_{c2,i}}\right\| \tag{14}$$

where $\mathbf{p}_{i,m}$ is 3-D coordinates calculated from corresponding feature points between camera position c_1 and c_2. $(u_{c1,i}, v_{c1,i})$ and $(u_{c2,i}, v_{c2,i})$ are image coordinates of feature points. The method calculates $a_{i,m}$ of all camera position combination. If the smallest value $a_{\min,i}$ is larger than a given threshold a_{th}, the feature point is rejected as a measurement result which has low accuracy.

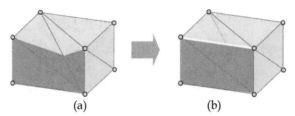

(a)   (b)

Fig. 11. Triangular mesh generation and its optimization. (a) Triangular meshes generated by Delaunay triangulation. (b) Optimized triangular meshes.

Measurement accuracy of straight-line is evaluated by equation (14), too. In this evaluation, $\mathbf{p}_{i,m}$ is the middle point of the line connecting two vectors $\mathbf{g}_i^{c1}$ and $\mathbf{g}_i^{c2}$ at the shortest distance. Image coordinates $(u_{c1,i}, v_{c1,i})$ and $(u_{c2,i}, v_{c2,i})$ are reprojection points of these vectors to images acquired at camera position c_1 and c_2.

5. Model construction

Triangular meshes are generated from integrated measurement data by using the 3-D Delaunay triangulation. However, Delaunay triangulation generates a triangular mesh which contradicts a physical shape because the triangular mesh does not consider the shape of the measurement object. Therefore, we apply the triangular optimization method (Nakatsuji et al., 2005) to the triangular mesh (Fig. 11). The method adapts the triangular mesh to the physical shape by detecting a texture distortion. By texture mapping to these meshes, a 3D environment model is constructed.

6. Experiments

First, accuracy of line-based measurement is evaluated. Measurement objects are lengthwise-lines on a flat wall shown in Fig. 12. The reason for including crosswise-lines is that the proposed method needs lines having different direction. The moving distance of the camera was about 2m. The number of input images is 72. An input image size is 2496×1664 pixels.

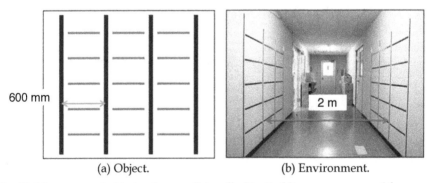

600 mm

2 m

(a) Object.   (b) Environment.

Fig. 12. Measurement objects place on flat walls. Vertical lines are measured for accuracy evaluation. Level lines are set for camera movement estimation.

Three-Dimensional Environment Modeling Based on Structure from Motion with Point and
Line Features by Using Omnidirectional Camera

75

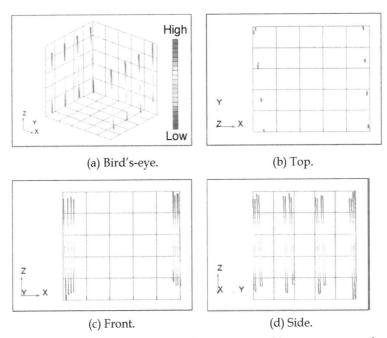

(a) Bird's-eye.   (b) Top.

(c) Front.   (d) Side.

Fig. 13. Measurement result for accuracy evaluation. Vertical lines are measured precisely.
Measurement results of level lines are removed as low measurement accuracy data.

	Angle (direction) [deg]	Angle (plane) [deg]	Depth [mm]
Standard deviation	1.2	-	2.3
Maximum	1.9	1.5	7.5

Table 1. Evaluation result of vertical line measurements.

A measurement result is shown in Fig. 13. A measurement result which has larger Z-coordinate value is displayed in red, and smaller one is displayed in blue. Angles and depth errors were calculated for evaluation of measurement accuracy in Table 1. An angle error of calculated line directions is 1.2 degree standard deviation. Its maximum error is 1.9 degree. An angle error between two flat walls estimated from measurement data is within 1.5 degrees. A depth error between an estimated flat wall and reconstructed lines has 2.3 mm standard deviation. Its maximum error is 7.5 mm. This experiment shows that our proposed method has sufficient accuracy to accomplish static obstacle avoidance, self-localization.

Next, we experimented in an environment including non-textured objects as shown in Fig. 14. We used 84 omnidirectional images. Measurement results of feature points and straight-lines are shown in Fig. 15. The blue marks in the figure show the camera trajectory. Although feature point measurement results are sparse, straight-lines can be measured densely. This experimental result shows that our proposed method is effective for a non-textured environment.

(a) Non-textured environment.

(b) Input image.

Fig. 14. Non-textured environment. We cannot get enough feature points in the environment because there are few features.

Modeling result is shown in Fig. 16. Images having a view-point which is different from camera observation points can be acquired. A model constructed from feature point measurement data is only a small part of this environment (Fig. 16(a) and (c)). Meanwhile, edge measurement data makes it possible to construct a non-textured environment model (Fig. 16}(b) and (d)).

We also experimented in an outdoor environment including textured objects (trees and so on) and non-textured objects (buildings and so on) as shown in Fig. 17. We used 240 omnidirectional images. An integrated measurement result is shown in Fig. 18. As one of textured objects, the shape of ground surface is measured by point-based measurement. As non-textured objects, the shape of the building is measured by line-based measurement.

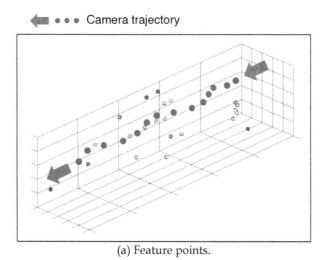

(a) Feature points.

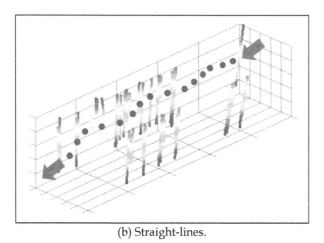

(b) Straight-lines.

Fig. 15. Measurement results of non-textured environment. (a) Although camera movement estimation is possible, we get sparse measurement results. (b) Straight-lines make it possible to measure non-textured environments densely.

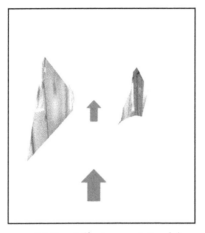 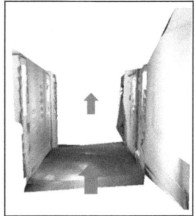

(a) Front (feature point only).   (b) Front (with straight-line).

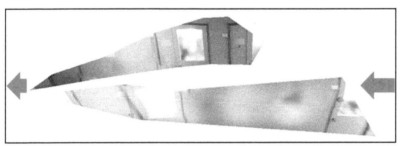

(c) Bird's-eye (feature point only).

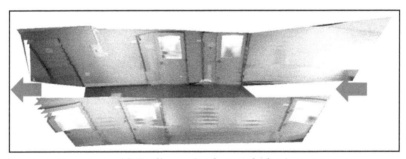

(d) Bird's-eye (with straight-line)

Fig. 16. Modeling results of non-textured environment. We can construction a model including many non-textured objects by the method with straight-lines.

Modeling result is shown in Fig. 19. By the combination of point-based measurement and line-based measurement, our proposed method can construct a model of 3-D environment including textured and non-textured objects. Experimental results showed the effectiveness of our proposed method.

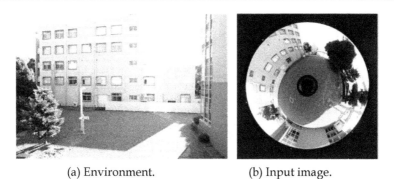

(a) Environment.   (b) Input image.

Fig. 17. Outdoor environment. There are textured objects (trees, tiles and so on) and non-textured objects (walls etc.)

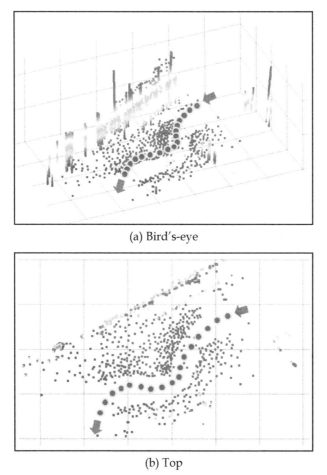

(a) Bird's-eye

(b) Top

Fig. 18. Measurement result of outdoor environment.

(a) Bird's-eye.

(b) Right.

(c) Front.

Fig. 19. Modeling results of outdoor environment.

7. Conclusions

We proposed an environment modeling method based on structure from motion using both feature points and straight-lines by using an omnidirectional camera. Experimental results showed that our proposed method is effective in environment including both textured and non-textured objects.

As future works, the precision improvement of edge tracking is necessary. Moreover, we should evaluate difference of camera movement estimation accuracy between point-based measurement and edge-based measurement. Further, edge position correlation should be used for increasing measurement stability.

8. References

Davison, A. J. (2003). Real-Time Simultaneous Localisation and Mapping with a Single Camera, *Proceedings of the 9th IEEE International Conference on Computer Vision*, Vol. 2, pp. 1403-1410, 2003.

Ishiguro, H. & Yamamoto, M. & Tsuji, S. (1992). Omni-Directional Stereo, *IEEE Transactions on Pattern Analysis and Machine Intelligence*, Vol. 14, No. 2, pp. 257-262, 1992.

Gluckman, J. & Nayar, S. K. (1998). Ego-motion and Omnidirectional Cameras, *Proceedings of the 6th International Conference on Computer Vision*, pp. 999-1005, 1998.

Bunschoten, R. & Krose, B. (2003). Robust Scene Reconstruction from an Omnidirectional Vision System, *IEEE Transactions on Robotics and Automation*, Vol. 19, No. 2, pp. 351-357, 2003.

Geyer, C. & Daniilidis, K. (2003). Omnidirectional Video, *The Visual Computer*, Vol. 19, No. 6, pp. 405-416, 2003.

Rachmielowski, A. & Cobzas, D. & Jagersand, M. (2006). Robust SSD Tracking with Incremental 3D Structure Estimation, *Proceedings of the 3rd Canadian Conference on Computer and Robot Vision*, pp. 1-8, 2006.

Kawanishi, R.; Yamashita, A. & Kaneko, T. (2009). Three-Dimensional Environment Model Construction from an Omnidirectional Image Sequence, *Journal of Robotics and Mechatronics*, Vol. 21, No. 5, pp. 574-582, 2009.

Chandraker, M. & Lim, J. & Kriegman, D. (2009). Moving in Stereo: Efficient Structure and Motion Using Lines, *Proceedings of the 12th IEEE International Conference on Computer Vision*, pp. 1741-1748, 2009.

Schindler, G.; Krishnamurthy, P. & Dellaert, F. (2006). Line-Based Structure from Motion for Urban Environments, *Proceedings of the 3rd International Symposium on 3D Data Processing, Visualization, and Transmission*, pp. 846-853, 2006.

Bartoli, A. & Sturm, P. (2005). Structure-from-motion using lines: representation, triangulation, and bundle adjustment, *Computer Vision and Image Understanding*, Vol. 100, Issue 3, pp. 416-441, 2005.

Smith, P. & Reid, I. & Davison, A. (2006). Real Time Monocular SLAM with Straight lines, *Proceedings of the 17th British Machine Vision Conference*, pp. 17-26, 2006.

Mariottini, G. L. & Prattichizzo, D. (2007). Uncalibrated video compass for mobile robots from paracatadioptric line images, *Proceedings of the 2007 IEEE/RSJ International Conference on Intelligent Robots and Systems*, pp. 226-231, 2007.

Spacek, L. A. (1986). Edge Detection and Motion Detection, *Image and Vision Computing*, Vol. 4, Issue 1, pp. 43-56, 1986.

Triggs, B. & McLauchlan, P. & Hartley, R. & Fitzgibbon, A. (1999). Bundle Adjustment -A Modern Synthesis, *Proceedings of the International Workshop on Vision Algorithms: Theory and Practice*, Springer-Verlag LNCS 1883, pp. 298-372, 1999.

Shi, J. & Tomasi, C. (1994). Good Features to Track, *Proceedings of the 1994 IEEE Computer Society Conference on Computer Vision and Pattern Recognition*, pp. 593-600, 1994.

Canny, J. F. (1986). A Computational Approach to Edge Detection, *IEEE Transactions on Pattern Analysis and Machine Intelligence*, Vol. PAMI-8, No. 6, pp. 679-698, 1986.

Fischler, M. A. & Bolles, R. C. (1981). Random Sample Consensus: A Paradigm for Model Fitting with Applications to Image Analysis and Automated Cartography, *Communications of the ACM*, Vol. 24, No. 6, pp. 381-395, 1981.

Nakayama, K. & Silverman, G. (1988). The Aperture Problem -- II. Spatial Integration of Velocity Information Along Contours, *Vision Research*, Vol. 28, No. 6, pp. 747-753, 1988.

Hartley, R. (1997). In defence of the eight-point algorithm, *IEEE Transactions on Pattern Analysis and Machine Intelligence*, Vol. 19, No. 6, pp. 580-593, 1997.

Nakatsuji, A. & Sugaya, Y. & Kanatani, K. (2005), Optimizing a Triangular Mesh for Shape Reconstruction from Images, *IEICE Transactions on Information and Systems*, Vol. E88-D, No. 10, pp. 2269-2276, 2005.

5

Vision Based Obstacle Avoidance Techniques

Mehmet Serdar Guzel and Robert Bicker
Newcastle University,
United Kingdom

1. Introduction

Vision is one of the most powerful and popular sensing method used for autonomous navigation. Compared with other on-board sensing techniques, vision based approaches to navigation continue to demand a lot of attention from the mobile robot research community. This is largely due to its ability to provide detailed information about the environment, which may not be available using combinations of other types of sensors. One of the key research problems in mobile robot navigation is the focus on obstacle avoidance methods. In order to cope this problem, most autonomous navigation systems rely on range data for obstacle detection. Ultrasonic sensors, laser rangefinders and stereo vision techniques are widely used for estimating the range data. However all of these have drawbacks. Ultrasonic sensors suffer from poor angular resolution. Laser range finders and stereo vision systems are quite expensive, and computational complexity of the stereo vision systems is another key challenge (Saitoh et al., 2009). In addition to their individual shortcomings, Range sensors are also unable to distinguish between different types of ground surfaces, such as they are not capable of differentiating between the sidewalk pavement and adjacent flat grassy areas. The computational complexity of the avoidance algorithms and the cost of the sensors are the most critical aspects for real time applications. Monocular vision based systems avoid these problems and are able to provide appropriate solution to the obstacle avoidance problem. There are two fundamental groups of vision based obstacle avoidance techniques; those that compute the apparent motion, and those that rely on the appearance of individual pixels for monocular vision based obstacle avoidance systems. First group is called as Optical flow based techniques, and the main idea behind this technique is to control the robot using optical flow, from which heading of the observer and time-to-contact values are obtained (Guzel & Bicker, 2010). One way of the control using these values is by acting to achieve a certain type of flow. For instance, to maintain ambient orientation, the type of Optic flow required is no flow at all. If some flow is detected, then the robot should change the forces produced by its effectors so as to minimize this flow, based on Law of Control (Contreras, 2007).

A second group is called Appearance Based methods rely on basic image processing techniques, and consist of detecting pixels different in appearance than that of the ground and classifying them as obstacles. The algorithm performs in real-time, provides a high-resolution obstacle image, and operates in a variety of environments (DeSouza & Kak, 2002). The main advantages of these two conventional methods are their ease of implementation and high availability for real time applications. However optical flow *based* methods suffer from two major problems, which are the illumination problem that varies with time and the problem of motion discontinuities induced by objects moving with respect to other objects or the background (Contreras, 2007). Various integrated methods for solving these problems have been proposed; nevertheless it is still a key challenge to employ optical flow for mobile robot navigation. Furthermore, appearance based methods also suffer from illumination problems and, are highly sensitive to floor stains, as well as to the physical structure of the terrain.

Consequently, while having significant performance advantages, there are certain drawbacks which restrict the applicability of these methods. In order to solve those challenges, a novel obstacle avoidance method is introduced in this chapter. The method is principally designed to fuse a Scale invariant features transform (SIFT) algorithm (Lowe, 1999), and template matching with a convolution mask technique, using a *Fuzzy Logic approach*. As opposed to the *Appearance based methods*, previously mentioned, an occupancy map of the environment is generated with respect to the local features and a template. The experimental results reveal that the proposed obstacle avoidance technique allows the robot to move efficiently within its environment and to successfully attain its local goals.

This chapter is organized as follows. In Section 2, the background knowledge to the conventional methods is briefly introduced. In Section 3, new technique is introduced. Section 4 provides the implementation of the behaviour-based robot and the experiment results from both the real and simulation experiments. Section 5 provides a summary of the work.

2. Background

In this section, optical flow based navigation techniques will first be outlined, followed by a brief introduction of the appearance based methods.

2.1 Optical flow

Optical flow, illustrated in Fig. 1, is an approximation to the motion field, summarizing the temporal change in an image sequence. The main idea behind the technique assumes that for a given scene point, the corresponding image point intensity I remain constant over time, which is referred as *conservation of image intensity* (Atcheson et al., 2009). Therefore, if two consecutive images have been obtained at the following time intervals, the basic idea is to detect the motion using image differencing. If any scene point projects onto image point (x, y) at time t and onto image point $(x + \delta x, y + \delta y)$ at time $(t + \delta t)$, the following equation is inferred based on the conservative of image intensity assumption.

$$I(x, y, t) = I(x + \delta x, y + \delta y, t + \delta t) \tag{1}$$

Expanding the right-hand side of the Eq. 1 using a Taylor series about (x, y, t), and ignoring the higher order terms then by rearrangement gives the following expression.

$$\delta x \frac{\partial I}{\partial x} + \delta y \frac{\partial I}{\partial y} + \delta t \frac{\partial I}{\partial t} = 0 \tag{2}$$

A simpler expression, is obtained by dividing by δt throughout and movement along the horizontal $\left(\frac{\delta x}{\delta t}\right)$, and vertical $\left(\frac{\delta y}{\delta t}\right)$ directions are u and v respectively. Having these rearrangements and denoting partial derivatives of I by I_x, I_y and I_t gives the differential flow equation shown in following expressions:

$$I_x u + I_y v + I_t = 0 \tag{3}$$

where, I_x, I_y and I_t are the partial derivatives of image brightness with respect to x, y and t, respectively. Having one equation in two unknowns δx, δy for each pixel is an aperture problem of the optical flow algorithms. To find the optical flow another set of equations is needed, using some additional constraint. All optical flow methods introduce additional conditions for estimating the actual flow. There are several methods employed to determine optical flow, namely: Block-based methods, differential methods, Phase Correlation and General variational methods (Atcheson et al., 2009). Differential methods are widely used for navigation tasks, and are mainly based on partial derivatives of the image signal and/or the sought flow field and higher-order partial derivatives. One of those methods is used to estimate flow vectors to steer the robots.

2.1.1 The optical flow method proposed by Horn and Schunk
Horn and Schunk proposed one of the most important optical flow methods using an gradient based approach (Horn & Schunck, 1981). According to their methodology, a regularizing term associated with smoothness is added to the general flow equation, as illustrated in Equation 3, in which neighbouring pixels have the same velocity as moving objects, so the brightness pattern of an image changes regularly. This constraint is demonstrated by minimizing the squares of gradient magnitudes. Smoothness of an optical flow area can also be calculated by determining the Laplacian of optical flow vectors speed both horizontal and vertical directions denoted by u and w respectively, illustrated in following expressions:

$$\nabla^2 u = \frac{\partial^2 u}{\partial x^2} + \frac{\partial^2 u}{\partial y^2}$$

$$\nabla^2 v = \frac{\partial^2 u}{\partial x^2} + \frac{\partial^2 u}{\partial y^2} \tag{4}$$

Where $E_i = \frac{\partial I}{\partial x}\delta x + \frac{\partial I}{\partial y}\delta y + \frac{\partial I}{\partial t}$ and $E_s = \nabla^2 u + \nabla^2 v$. The aim is to minimize the total error given by the following expressions that includes σ as the regularization parameter, controlling the association between the detail and the smoothness. High values of σ makes the smoothness constraint dominate and leads to a smoother flow

$$\iint \left(E_i^2 + \sigma^2 E_s^2 \right) dxdy \tag{5}$$

Horn and Schunk, can be used as the main reference to understand and solve the given error function, from which a pair of equations for each point can be obtained. Direct solution of these equations such as Gaus-Jordan Elimination (Bogacki, 2005) would be very costly. Instead, an iterative Gauss Seidel, approach is used to reduce the cost and obtain the flow vectors, as follows (Horn & Schunck, 1981):

$$u^{n+1} = \bar{u}^n - \left(\frac{I_x \bar{u}^n + I_y \bar{u}^n + I_t}{\sigma^2 + I_x^2 + I_y^2} \right); v^{n+1} = \bar{v}^n - \left(\frac{I_x \bar{v}^n + I_y \bar{v}^n + I_t}{\sigma^2 + I_x^2 + I_y^2} \right) \tag{6}$$

where I_x I_y and I_t are the partial derivatives with respect to x,y and t respectively, and the superscript $n+1$ denotes the next iteration, which is to be calculated and n is the last calculated result (Horn & Schunck, 1981).

Fig. 1. Optical Flow vectors (Guzel & Bicker, 2010).

2.2 Optical flow for mobile robot navigation

Flow vectors are utilized to navigate autonomous systems based on the Balance Strategy (Souhila & Karim, 2007), shown in the following equation, and the depth information which is extracted from the image sequence using Focus of Expansion(FOE) and Time To Contact values(TTC) (Souhila & Karim, 2007). The fundamental idea behind the Balance strategy is that of motion parallax, when the agent is translating, closer objects give rise to faster motion across the retina than farther objects. It also takes advantage of perspective in that closer objects also take up more of the field of view, biasing the average towards their associated flow (Contreras, 2007). The agent turns away from the side of greater flow. This control law is formulated by:

$$\Delta(F_l - F_r) = \left(\frac{\sum|w_L| - \sum|w_R|}{\sum|w_L| + \sum|w_R|}\right) \tag{7}$$

where $\sum|w_L|$ and $\sum|w_R|$ are the sum of the magnitudes of optical flow in the visual hemi fields on both sides of the robot's body. The following expression gives the new heading angle

$$\theta_{new} = (\Delta(F_l - F_r) \times k) \tag{8}$$

where k, is a constant, and used to convert the obtained result to an appropriate control parameter to steer the robot.

2.3 Appearance-based methods

Appearance based methods that identify locations on the basis of sensory similarities are a promising possible solution to mobile robot navigation. The main idea behind the strategy is to head the robot towards the obstacle-free position using similarities between the template and the active images (F. Vassallo et al., 2000). The similarity between the image patterns can be obtained by using feature detectors, involving corner based detectors, region based detectors and distribution based descriptors (Alper et al., 2006). However, most of these techniques consume a lot of process on time which is not appropriate for real time systems. In order to handle this problem in mobile robot applications, algorithms are designed based on the appearance of individual pixels. The classification of the obstacles is carried out by using the pixel difference between the template and active image patterns.In principally; any pixel that differs in appearance from the ground is classified as an obstacle. However, the method requires three assumptions that are reasonable for a variety of indoor and outdoor environments which are (Saitoh et al., 2009):

a.   Obstacles must be different in appearance from the ground.
b.   The ground must be flat.
c.   There must be no overhanging obstacles.

The first assumption is to distinguish obstacles from the ground, while the second and third assumptions are required to estimate the distances between detected obstacles and the robot. There are several models for representing colour. The main model is the RGB(Red, Green, Blue) model which is used in monitor screens and most image file formats however, colour information for RGB model is very noisy at low Intensity. The RGB format is mostly converted to a HSV (Hue, Saturation, and Value). In HSV, Hue is what humans perceive as colour, S is saturation and Value is related to brightness, (or HIS(Hue, Intensity, Saturation) model) and in HIS, H and S represents the same as parameters in HSV colour models but I is an intensity value with a range between [0,1] where 0 is black and white is 1. These colour spaces are assumed to be less sensitive to noise and lighting conditions. The flow chart of the appearance based obstacle detection systems is illustrated in Figure 2. The input image is first convolved with a smoothing filter to reduce the noise effects, and then smoothed image is converted to *HIS, HSV* or any related *colour space* with respect to the developed algorithm (Fazl-Ersi & Tsotsos, 2009). A reference area is obtained from this image which might be any shape of geometry such as trapezoidal, triangle or square, and histogram values of this reference area are generated (Saitoh et al., 2009). Finally, a comparison between the reference image and the current image is made using some predefined threshold values. For instance, assume that the bin value, Hist($H(x, y)$), of the generated histogram and the threshold value,T_H, are compared, where $H(x, y)$ is the H value at pixel (x, y). If Hist($H(x, y)$) > T_H then the pixel (x, y) is classified into the safe region, or else it is classified into the obstacle region. In order to simply the problem, the results are represented in a binary image in which the safe path is represented with white but the obstacles are represented with black, as illustrated in Figure 3.However, identifying places purely on the basis of sensory similarity is too simplistic; different places may look very similar, even with a rich sensing methodology due to lighting conditions, shadows on illumination Furthermore, for dynamic environments there might be unexpected stains on the ground which may be the detected as an obstacle and leads the robot to an unsafe path. An example with respect to this case is illustrated in Figure 4.

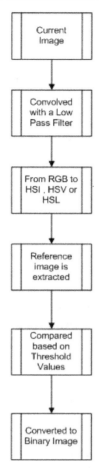

Fig. 2. Flow chart of the Appearance Based obstacle detection algorithm.

Fig. 3. Appearance based obstacle detection method.

Fig. 4. Effects of lighting conditions and unexpected stains on the floor.

3. SIFT and template matching based obstacle avoidance strategy

In order to cope with the drawbacks of the conventional appearance based methods, a novel feature matching based technique, comprising a Scale Invariant Feature Transform and Template matching with a convolution mask, will be discussed in this section. The detail of the control algorithms with respect to these techniques is illustrated in Figure 5. Before introducing the proposed control algorithm and fusion technique, essential background knowledge regarding the SIFT and Template matching will be presented.

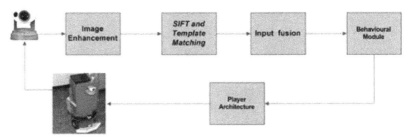

Fig. 5. Control architecture of the obstacle avoidance system.

3.1 Scale invariant feature transform (SIFT)

The Scale Invariant Feature Transform formerly abbreviated as SIFT is an algorithm in computer vision to detect and describe local features in images. The algorithm was published by Lowe (Lowe, 1999a, 2004b), and since then has been accepted as one of the most powerful local feature detection technique. The most notable improvements provided by SIFT are invariance to scale and rotation, and accuracy in feature point localization and matching. The evaluations carried out proposes that SIFT-based descriptors, which are region-based, are the most robust and distinctive, and are therefore best suited for feature matching. A summary of the SIFT methodology is illustrated in Fig. 6. (Lowe, 1999a, 2004b). The initial state of this algorithm is Scale space extreme detection where the interest points, which are called key-points in the SIFT framework, are detected. For this, the image is convolved using Gaussian filters, proved the only possible scale-space kernel, at different

Problem	Technique	Advantage
key localization / scale / rotation	DoG / scale - space pyramid / orientation assignment	accuracy, stability, scale & rotational invariance
geometric distortion	blurring / resampling of local image orientation planes	affine invariance
indexing and matching	nearest neighbour / Best Bin First search	Efficiency / speed
Cluster identification	Hough Transform voting	reliable pose models
Model verification / outlier detection	Linear least squares	better error tolerance with fewer matches
Hypothesis acceptance	Bayesian Probability analysis	reliability

Fig. 6. SIFT Methodology (Lowe, 1999a, 2004b).

scales, and then the difference of successive Gaussian-blurred images are obtained, illustrated in Figure 7. The convolved images are grouped by octave which corresponds to doubling the value of standard deviation of the Gaussian distribution (σ).The Convolution of the image at scale $k\sigma$ with a Gaussian filter is expressed as follows:

$$L(x, y, k\sigma) = G(x, y, k\sigma) * I(x, y) \tag{9}$$

where,

$$G(x, y, \sigma) = \frac{1}{(2\pi\sigma^2)} exp^{-(x^2+y^2)/2\sigma^2} \tag{10}$$

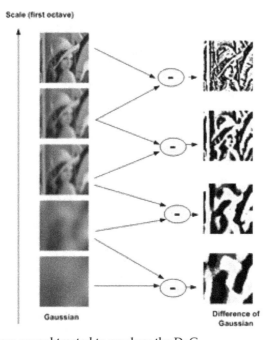

Fig. 7. Gaussian images are subtracted to produce the DoG.

When DoG images have been obtained, key- points are identified as local minima/maxima of the DoG images across scales. This is done by comparing each pixel in the DoG images to its eight neighbors at the same scale and nine corresponding neighboring pixels in each of the neighboring scales. If the pixel value is the maximum or minimum among all compared pixels, it is selected as a candidate keypoint', as shown in Figure 8.

Following steps are Key-point localization and Orientation assignment (Lowe, 1999a, 2004b). After key-point orientation has been completed, each key specifies stable 2D coordinates, comprising x, y, scale and orientation. Finally, a signature, local descriptor, is computed as a set of orientation histograms on 4x4 pixel neighbourhoods. Histograms have 8 bins each, and each descriptor contains an array of 4 histograms around the key-point. This leads to a SIFT feature vector with 8x4x4 = 128 elements, illustrated in Figure 9. This vector is normalized to enhance invariance to changes in illumination.

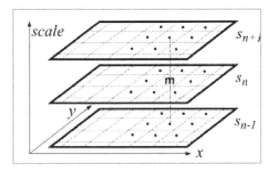

Fig. 8. A key-point is defined as any value in the DoG.

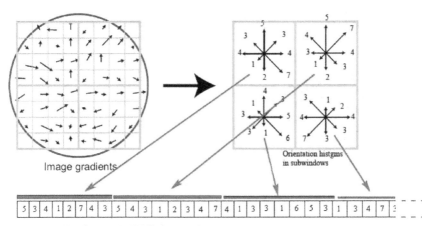

Fig. 9. SIFT Feature Descriptor.

3.1.1 SIFT matching

Feature vectors which extracted from the SIFT algorithm to solve common computer vision problems, comprising object detection, 3D scene modeling, recognition and tracking, robot localization and mapping. This procedure requires an appropriate and fast matching algorithm. An example with respect to the SIFT matching is illustrated in Figure 10. The main matching algorithm is able to find each key-point by identifying its nearest neighbor in the database of key-points from training images. The nearest neighbor is defined as the key-point with minimum Euclidean distance for the invariant descriptor vector, as previously discussed.However, many features may not be matched correctly due to background clutter in natural images or may not have correct match in the training database, and hence, mismatches should be discarded in order to obtain accurate results. Global thresholding on distance to the closest feature does not perform well, as some descriptors are much more discriminative than others. Alternatively, a more effective measure is obtained by comparing the distance of the closest neighbor to that of the second-closest neighbor, which performs well. An appropriate threshold value regarding this comparison, called distance ratio, is employed to reject false matches while increasing the correct matches. This value varies from 0.1 to 0.9, depending on the application type. In this case, 0.7 is employed with respect to the experimental results, which eliminates %90 of the false matches while discarding almost %10 correct matches. In addition, to reject rest of all false matches, an essential statistical method is applied to the matching space, fundamentally using the rate between scale and orientation parameters of the feature vectors. According to this method, even though the matching is validated between any two feature vectors, if the scale or the orientation parameter rate between them is more than a threshold value, matching is discarded; this procedure performs robustly to decrease false matches over all data sets (Lowe, 1999a,2004b).

Fig. 10. SIFT matching.

3.2 Template matching

Template matching is a simple and popular technique in computer vision and image processing to find small parts of an image which match a template image. It can be used in mobile robot navigation or as a way to detect edges or objects in images; an example with respect to this technique is illustrated in Figure 11. A basic method of template matching uses a convolution mask which can be easily performed on grey images. The convolution output will be the highest at places where the image structure matches the mask structure, i.e. where large image values get multiplied by large mask values. This method is normally implemented by first picking out a part of the search image to use as a template.

Fig. 11. Template Matching.

For instance, the input and output images are called I(x, y) and O(x, y) respectively, where (x, y) represent the coordinates of each pixel in the images and the template is called $T(x_t, y_t)$, where (x_t, y_t) represent the coordinates of each pixel in the template. The technique simply moves the centre of the template $T(x_t, y_t)$ over each (x, y) point in the search image and calculates the sum of products between the coefficients in I(x, y) and $T(x_t, y_t)$ over the whole area spanned by the template. As all possible positions of the template with respect to the Input image are considered, the position with the highest score is the best position, and which is represented in the output image. There are several techniques to handle translation problem; these include using SSD (Sum of squared differences), CC (Cross Correlation) and SAD (Sum of absolute differences) (Wen-Chia & Chin-Hsing, 2009). One of the most powerful and accurate of those is CC, which basically measures the similarity of two variables and defined as follows (9):

$$Cor = \frac{\sum_{i=0}^{N-1}(x_i - \dot{x}) \times (y_i - \dot{y})}{\sqrt{\sum_{i=0}^{N-1}(x_i - \dot{x})^2 \times \sum_{i=0}^{N-1}(y_i - \dot{y})^2}} \tag{11}$$

Where N is the template image size; $\dot{x}$ and $\dot{y}$ represents average gray level in the template and source image respectively. The goal is to find the corresponding (correlated) pixel within a certain disparity range that minimizes the associated error and maximizes the similarity. This matching process involves computation of the similarity measure for each disparity value, followed by an aggregation and optimization step (ZitovÃ¡ & Flusser, 2003). An example related to correlation based technique is illustrated in Figure 12.

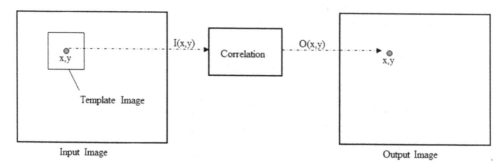

Fig. 12. Correlation technique based Template Matching.

3.3 Obstacle avoidance using SIFT and appearance based

Appearance based methods have significant processing and performance advantages which make them a good alternative for vision based obstacle avoidance problems. However as mentioned previously, there are certain drawbacks, which *restricts the applicability of these methods*. In order to handle these drawbacks, the results of the conventional method is improved by using the results of *SIFT* based feature matching approach. The flowchart diagram of the proposed algorithm is illustrated in Figure 13. First the acquired image is smoothed using a Gaussian filter to eliminate the noise in the image. Then a copy of the original image is converted to PGM image format which is required to carry out the *SIFT* matching process. Both images are divided into 16 sub-images composed of 44×36 pixels. For each sub images, template matching using cross correlation and *SIFT* matching are performed against reference images, illustrating the safe route, simultaneously. The results for each segment are fused using fuzzy logic to build up a sufficiently accurate occupation map of the environment.

The robot, which employs a Subsumption Architecture (Brooks, 1986), is successfully directed along a collision-free path using this map. To evaluate the performance of the proposed algorithm, it is applied to a test case, as illustrated in Figure 14. The matching results, shown in Figure 15, indicate that both techniques generate similar results under ideal conditions, involving low illumination changes and flat ground. However, template matching may fail to provide accurate matching results against illumination problem, and non-flat surfaces. In order to handle these cases, *SIFT* matching provides a reliable matching strategy, which is able to match the extracted features, invariant to scale, orientation, affine distortion, and partially invariant to illumination changes, with high accuracy. An example illustrating this condition can be seen in Figure 16, comprising 9th and 10th sub-image sequences. Despite the illumination problem and non-flat ground, the SIFT matching performs well for these cases illustrated in Figure 17. Consequently; the results indicate that instead of using each method separately, fusion of them generates more reliable results. A fuzzy logic based approach with respect to this fusion procedure will be discussed in the following section.

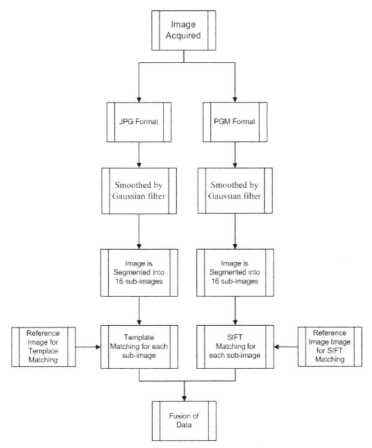

Fig. 13. An example from the testing environment.

Fig. 14. An example from the testing environment.

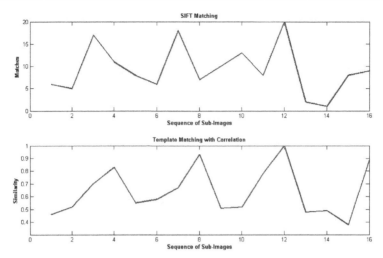

Fig. 15. Comparison of two algorithms.

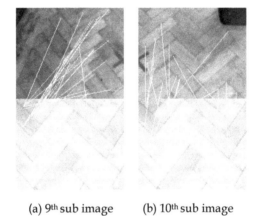

(a) 9th sub image   (b) 10th sub image

Fig. 16. Matching results of two algorithms.

Fig. 17. False matching generated by SIFT matching.

3.4 Fuzzy logic

Fuzzy logic (FL) is a form of many-valued logic derived from fuzzy set theory to deal with reasoning that is robust and approximate rather than brittle and exact. In contrast with two-valued Boolean logic, FL deals with degrees of membership and degrees of truth. FL uses the continuum of logical values between 0 (completely false) and 1 (completely true). FL has been utilized as a problem-solving control system by several researchers for different problems. Since, FL lends itself to implementation in systems ranging from simple, small, embedded micro-controllers to large, networked, multi-channel PC or workstation-based data acquisition and control systems.

FL is fundamentally easy to implement and provide faster and more consistent results than conventional control methods. In this study, a FL based control system based on the Mamdani method is designed to fuse given algorithms. The basic configuration of a fuzzy-logic system is composed of three parts: *Fuzzification*, *Inference Mechanism* and *Deffuzification (Driankov, 1987)*. These will be presented and associated with the fusion problem in the following parts.

3.4.1 Fuzzification

Fuzzfication comprises a scale of transformation of input data of a current process into a normalised domain. This process requires the identification of two parts: the first part defines the fuzzy variables that correspond to the system input variables. The second part is to define the fuzzy sets of the input variables and their representative membership functions including the ranges of the data. Membership function, may cross the boundary of another fuzzy membership function. Each membership function may be triangular, a trapezoidal or bell shaped, as illustrated in Figure 18. The choice of the fuzzy sets is based on expert opinion using natural language terms that describe the fuzzy values. In this study triangle and trapezoid models are utilized to design membership functions of input and output values.

Fuzzy logic uses intersection, union, and complement operations to represent the standard common operators of AND, OR, and NOT, respectively. The most common method used to calculate intersection and union operations are the Minimum and Maximum functions. For the fuzzy sets M and N which are subsets of the universe X, the following definitions are proposed to represent the AND, OR, and NOT operators, respectively (Ross & Hoboken, 2004) (see Figure 19).

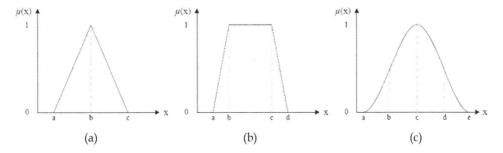

Fig. 18. Membership function shapes, (a) Triangular, (b) Trapezoidal, (c) Gaussian.

The triangular function has three parameters which can be defined as follows:

$$\mu(x) = \begin{cases} 0, & x < a \\ \frac{x-a}{b-a}, & a \leq x < b \\ \frac{c-x}{c-b}, & b \leq x \leq c \\ 0, & x > c \end{cases} \quad (12)$$

The trapezoidal function incorporates four parameters can be represented as:

$$\mu(x) = \begin{cases} 0, & x < a \\ \frac{x-a}{b-a}, & a \leq x < b \\ 1, & b \leq x \leq c \\ \frac{d-x}{d-c}, & c < x \leq d \\ 0, & x > d \end{cases} \quad (13)$$

3.4.2 Inference mechanism

The generation of the fuzzy rules is a first step which depends on the knowledge and experience of the human operators, the fuzzy model of the plant concerned, and an analysis of the system. The rule-base is composed of two parts namely, the IF-part and the THEN-part. The IF-part is the antecedent part where rules are defined to describe the system state in terms of a combination of fuzzy propositions while the THEN-part is the consequent part which forms the desired conclusion of the output variable. Afterwards, fuzzy inference provides the conclusion of the rule-base and forms the intermediate stage between the fuzzification and defuzzification of the fuzzy system. There are two methods used to find the rules conclusion namely Max-Min inference and Max-Product inference. Max-Min utilizes the Minimum operator to combine the antecedent of the IF-THEN rules which produces modified fuzzy sets for the outputs. These modified sets are then combined using the Maximum operator and Max-Product inference and utilizes the standard Product operator to combine the antecedent of the IF-THEN rules. Then the Maximum operator is used to combine these modified sets (Ross & Hoboken, 2004).

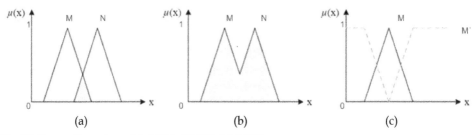

Fig. 19. Fuzzy Operators, (a) AND, (b) OR, (c) NOT

3.4.3 Defuzzification

Defuzzification is the process of mapping from a space of inferred fuzzy control action to a space of non-fuzzy control actions where the calculated crisp value is that which best represents the inferred control action. Several methods can be used to calculate this crisp value such as the Centre-of-Area, Centre-of-Largest-Area, Centre-of-Sums, and Mean-of-Maximum. These methods are based on two basic mechanisms: Centroid and Maximum. The centroid methods are based on finding a balance point while the Maximum methods search for the highest peak of weight (area) of each fuzzy set (Ross & Hoboken, 2004). Centre-of-Sum is used in this study which is faster than many defuzzification methods, and is not restricted to symmetric membership functions. This process performs the algebraic sum of individual output fuzzy sets instead of their union, illustrated in the following equation:

$$\mu^* = \frac{\int y \sum_{n=1}^{k} \mu_n(y) d_y}{\int \sum_{n=1}^{k} \mu_n(y) d_y} \tag{14}$$

3.4.4 Fusion of algorithms with fuzzy logic

The fundamental architecture of the proposed system is illustrated in Figure 20. The FL controller has two inputs, namely SIFT and Correlation which involves matching strength and similarity rate respectively. The output of the controller generates an appropriate turning rate (w) to avoid obstacle. In order to adapt the results of vision based algorithms efficiently as well as to operate the robot smoothly, the image is divided into n clusters (sub-images), based on the resolution of the image, its parts, and for each cluster a fuzzy fusion algorithm is applied. SIFT matching value of each cluster is rescaled by multiplying with n which normalizes the input for Fuzzy Inference System (FIS). Final turning rate is calculated to sum up all clusters, considering the sign of each part which is defined as follows:

$$w = \sum_{i=1}^{n} w_i, \; n \text{ is even, left clusters are positive and right ones are negative} \tag{15}$$

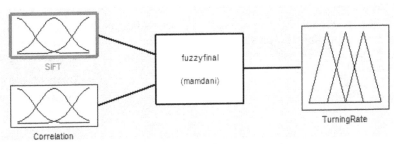

Fig. 20. Architecture of Fuzzy control system for fusing operation.

To generate a value of turning rate (w), a Max-Min type FIS and a Centre-of-Sums defuzification method were used, as previously discussed. The first step is to design membership functions for fuzzification and defuzzification processes. While, several researchers utilize different membership function shapes regarding to the problem in various applications. The trapezoidal and triangular shapes have been selected in this work to simplify the computation. However, it should be noted that there are no precise methods to adjust the membership functions. Table 1 provides an example of the selection of the fuzzy terms that describe linguistic variables used in this study. For this study three membership functions are defined *namely: SIFT (s_m), Correlation (c_s) and Turning Rate (w).* SIFT function, representing the matching strength between the reference image and the current image, is illustrated in Figure 21(a). Whereas the correlation function represents the similarity between the reference and current images, as illustrated in Figure 21(b). Turning Rate is the output function which represents the angular velocity value to steer the robot whilst avoiding obstacles (see Figure 22). The next step is to design appropriate fuzzy rules depending on the detail of each Fuzzy Inference System (FIS). A set of experiments were carried out until the outputs are judged to satisfy each different situations. Table 2 displays the fuzzy rules for the given problem.

Linguistic Variables	Linguistic Terms
SIFT	Weak, Medium, Strong
Correlation	Weak, Medium, Strong
Turning Rate	Straight, Less, Medium, Sharp

Table 1. Linguistics variables and their linguistics terms.

Inputs	Correlation (c$_s$)		
SIFT (s$_m$)	Weak	Medium	Strong
Weak	Sharp	Medium	Less
Medium	Sharp	Medium	Straight
Strong	Medium	Less	Straight

Table 2. Fuzzy rule-base for Turning rate (w).

Test results lead to tune the system by changing rules, adjusting the membership functions shapes of both input and outputs. Once the procedure has been run several times, a consistent system is attained. The following section will integrate the proposed algorithm to a behavioral based architecture.

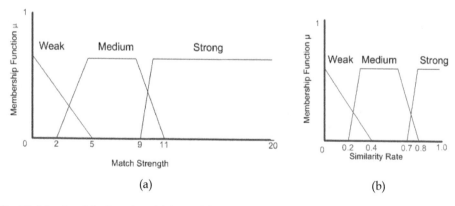

(a)   (b)

Fig. 21. Membership function, (a) 'input' SIFT matching, (b) 'input' Correlation.

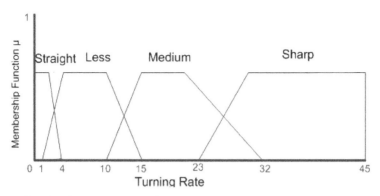

Fig. 22. Membership function, 'output' Turning rate.

4. Evaluation and implementation of the proposed system

In order to evaluate the performance of the algorithm, it is integrated using a behavioral based architecture. There are several approaches to designing a behavioural-based architecture depending on the required task. In this study, the architecture has been designed based on the subsumption architecture in which each layer or behaviour implements a particular goal of the robot and higher layers are increasingly abstract. Each layer's goal subsumes that of the underlying layer, and their interaction with each other will be illustrated by using finite state machines (FSM) which defines several states (behaviours) that represents a current situation for the robot. Certain events from the outside of the world can change the state. For instance, the robot could have a *Goto* state whereby it is moving about the environment trying to get closer to its goal. When any obstacle is detected nearby, the state may change from *Goto* to *Obstacle Avoidance*, and the avoidance algorithm will move the robot away from the obstacle. When the obstacle has been avoided, the robot state will change back to the *Goto*. The architecture, designed for this study, comprises three behaviors, namely: *Goto, Avoid_ Obstacle* and *Finish*. *Goto* behavior steers the robot to a specific goal position, *Avoid_Obstacle behaviour* utilizes the proposed vision based intelligent algorithm to avoid obstacles, and *Finish* behavior is merely enabled after the goal is found, and the robot is stopped. FSM diagram of the system is illustrated in Figure 23.

The system was developed using a Pioneer 3-DX mobile robot, with an on-board IntelPentium 1.8 GHz (Mobile) processor, and includes 256 Mbytes of RAM memory, as shown in Figure 24 (a). The mobile robot used in this study has been developed as a part of the Intelligent Robot Swarm for Attendance, Recognition, Cleaning and Delivery (IWARD) project. An Axis-213 camera, 25 frame rate, was integrated into this system. The software architecture of the proposed system is supported by CIMG Library and Player Architecture, which are open-source software projects. All experiments were conducted in an area of the Robotics and Automation Research Laboratory of Newcastle University, which has physical dimensions of 15.60m x 17.55m, as illustrated in Figure 24 (b). The camera tilted down 30 degrees to detect the floor precisely, and a reference image was taken from this environment. To evaluate the performance of the system, several different scenarios were performed and four of them will be discussed in this section.

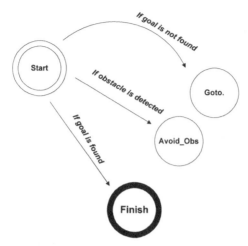

Fig. 23. FSM diagram of the behavioral architecture.

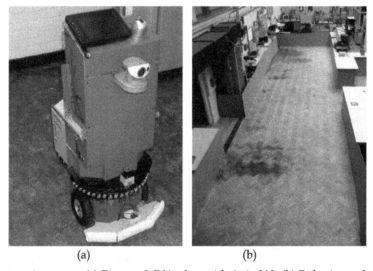

(a)   (b)

Fig. 24. Test environment, (a) Pioneer 3-DX robot with Axis-213, (b) Robotics and Automation Laboratory

Scenario 1: The mobile robot is required to navigate from 'Start Position' (-6, 0) along a forward direction in a partially cluttered enviorement and avoids one obstacle located along its path as shown in Figure 25 (a). Performance of the image processing algorithms during the task is given Figure 25 (b). The robot's GoTo behaviour steers the robot forward direction.

Scenario 2: The mobile robot navigates from 'Start Position' (-6, 0) along a forward direction in a partially cluttered enviorement and has to avoid two obstacles located along its path as shown in Figure 26 (a) and evaluation of the image processing algorithms illustrated in Figure 26 (b). The robot's GoTo behaviour steers the robot forward direction.

Scenario 3: The mobile robot navigates from 'Start Position' (-7, 0) along a forward direction, two obstacles located on both sides of its path as shown in Figure 27 (a). Performance of the image processing algorithms are illustrated in Figure 27 (b). The robot's GoTo behaviour steers the robot forward direction.

Scenario 4: The mobile robot is required to navigate from 'Start Position' (-6, 0) to the 'Goal Position' (3, 0.5) in a partially cluttered enviorement as shown in Figure 28(a). Performance of the image processing algorithms during the task is given Figure 28 (b). The robot's GoTo behaviour steers the robot to a specific position which evaluates the performance of the obstacle avoidance algortihm with a Wavefront path planning algorithm (Barraquand & Latombe, 1991).

Figure 25 (a) presents the navigation results for scenario 1, in which the robot steers forward until it percives the obstacle. The robot avoids the obstacle succesfully and then keeps going forward until it detects the wall, after which it avoids the wall and continues heading forward until it encounters the door. The robot succesfully avoids the door and continues moving.

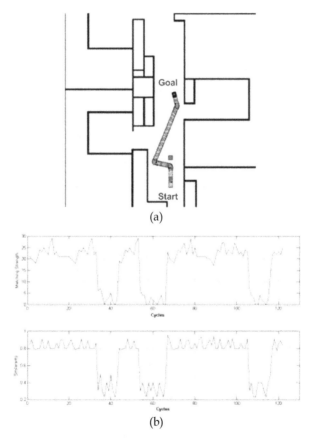

(a)

(b)

Fig. 25. Scenario 1, (a) Estimated Trajectory, (b) Performance of the image processing algorithms.

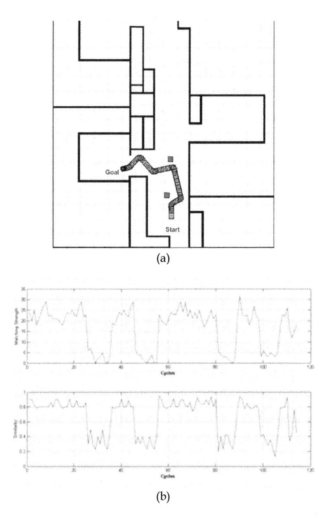

(a)

(b)

Fig. 26. Scenario 2, (a) Estimated trajectory, (b) Performance of the image processing algorithms.

Figure 26(a) displays the test results for scenario 2. Two obstacles are located along the robot's path in the forward direction. The robot navigates until it detects the first obstacle, then it avoids the obstacle and steers forward. Having detected the wall, the robot ceases forward motion and avoids the wall, after which, it moves forward until it detects the second obstacle, which it avoids succesfully. Subsequently, the robot avoids the door and the wall respectively.It finally moves forward until it is ceased. The third simulation is given Figure 27(a) where the robot attempts to move forward while passing the gap between two obstacles. It keeps moving forward until the obstacles are detected, initially turns left, followed by a right maneuver. After the robot succesfully avoids the objects it then resumes its path and continues moving forward.

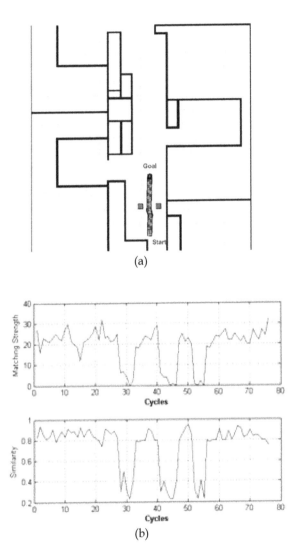

(a)

(b)

Fig. 27. Scenario 3, (a) Estimated trajectory, (b) Performance of the image processing algorithms.

Figure 28(a) displays the test results for scenario 4. The aim of this scenario is to test the avoidance strategy with a Wavefront path planning algortihm (Barraquand & Latombe, 1991) provided by the Player Architecutre. The robot navigates towards the 'first waypoint' until it perceves the obstacle. The robot avoids the obstacle and resumes its desired path. As there are no obstacles located along the rest of its path to the goal, the robot, therefore navigates directly to its goal.

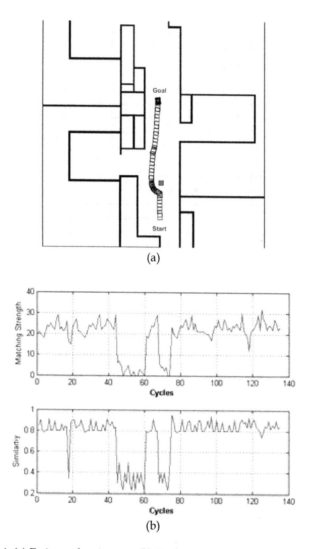

Fig. 28. Scenario 4, (a) Estimated trajectory, (b) Performance of the image processing algorithms.

5. Conclusion

The aim of this research was that it should be possible to develop a robust intelligent vision based obstacle avoidance system that can be adapted to realistic navigation scenarios. Most of the previously proposed techniques suffer from many problems. One of the most popular of those is *Apperance-based* methods which basically consist of detecting pixels different in appearance than the ground and classifying them as obstacles. Conventional *Appearance-*

based methods utilize simple template matching techniques which are fast but highly sensitive to lighting conditions and structure of the terrains. In addition, *Optical flow-based* methodologies,, the pattern of apparent motion of objects in a visual scene caused by the relative motion relying on apparent motion, are also highly sensitive to lighting conditions and suffer from nose.

To overcome those problems, a new obstacle avoidance method has been proposed which is inspired from feature matching principal. *SIFT-based* descriptors outperform other local descriptors on both textured and structured scenes, with the difference in performance larger on the textured scene. Accordingly, conventional *SIFT* algorithm, which is able to perform with high accuracy and reasonable processing time, is employed to match images, and it is finally adapted to an obstacle avoidance technique. According to the techniques, reference image, presenting the free path, is matched with the current image during the navigation to estimate the steering direction. However, preliminary experimental results reveals that despite the *SIFT* algorithm performs better than conventional methods with stained environment and compensates lighting failures, they may fail or produce several mismatched features due to low resolution or vibration caused by navigation over rough terrains. Therefore instead of using each method individually, conventional template matching method and *SIFT-based* descriptor are fused by using an appropriate *Fuzzy Fusion* algorithm which increases the accuracy and compensates the errors caused by lighting conditions and stains.

In order to verify the performance of the proposed obstacle avoidance algorithm, Real experiments to guide a *Pioneer 3-DX mobile robot* in a partially cluttered environment are presented, Results validate that proposed method provides an alternative and robust solution for mobile robots using a single low-cost camera as the only sensor to avoid obstacles.

6. References

Alper, Y., Omar, J. & Mubarak, S. (2006) Object tracking: A survey. *ACM Comput. Surv.*, 38, 13.

Atcheson, B., Heidrich, W. & Ihrke, I. (2009) An evaluation of optical flow algorithms for background oriented schlieren imaging. *Experiments in Fluids*, 46, 467-476.

BArraquand, J. & Latombe, J.-C. (1991) Robot motion planning. A distributed representation approach. *International Journal of Robotics Research*, 10, 628-649.

Bogacki, P. (2005) HINGES - An illustration of Gauss-Jordan reduction. *Journal of Online Mathematics and its Applications*.

Brooks, R. A. (1986) ROBUST LAYERED CONTROL SYSTEM FOR A MOBILE ROBOT. *IEEE journal of robotics and automation*, RA-2, 14-23.

Contreras, E. B. (2007) A biologically inspired solution for an evolved simulated agent. *Proceedings of GECCO 2007: Genetic and Evolutionary Computation Conference*. London.

Desouza, G. N. & Kak, A. C. (2002) Vision for mobile robot navigation: A survey. *IEEE Transactions on Pattern Analysis and Machine Intelligence*, 24, 237-267.

Driankov, D. (1987) Inference with a single fuzzy conditional proposition. *Fuzzy Sets and Systems*, 24, 51-63.

F. Vassallo, R., Schneebeli, H. J. & Santos-Victor, J. (2000) Visual servoing and appearance for navigation. *Robotics and Autonomous Systems, 31,* 87-97.

Fazl-Ersi, E. & Tsotsos, J. K. (2009) Region classification for robust floor detection in indoor environments. *Lecture Notes in Computer Science (including subseries Lecture Notes in Artificial Intelligence and Lecture Notes in Bioinformatics).* Halifax, NS.

Guzel, M. S. & Bicker, R. (2010) Optical flow based system design for mobile robots. *2010 IEEE Conference on Robotics, Automation and Mechatronics, RAM 2010.* Singapore.

Horn, B. K. P. & Schunck, B. G. (1981) Determining optical flow. *Artificial Intelligence, 17,* 185-203.

Lowe, D. G. (1999) Object recognition from local scale-invariant features. *Proceedings of the IEEE International Conference on Computer Vision.* Kerkyra, Greece, IEEE.

Ross, T. J. & Hoboken, N. J. (2004) *Fuzzy Logic with engineering applications* Hoboken, NJ:Wiley.

Saitoh, T., Tada, N. & Konishi, R. (2009) Indoor mobile robot navigation by central following based on monocular vision. *IEEJ Transactions on Electronics, Information and Systems,* 129, 1576-1584+18.

Souhila, K. & Karim, A. (2007) Optical flow based robot obstacle avoidance. *International Journal of Advanced Robotic Systems,* 4, 13-16.

Wen-Chia, L. & Chin-Hsing, C. (2009) A Fast Template Matching Method for Rotation Invariance Using Two-Stage Process. *Intelligent Information Hiding and Multimedia Signal Processing, 2009. IIH-MSP '09. Fifth International Conference on.*

Zitovã¡, B. & Flusser, J. (2003) Image registration methods: A survey. *Image and Vision Computing,* 21, 977-1000.

6

Non-Rigid Obstacle Avoidance for Mobile Robots

Junghee Park and Jeong S. Choi
Korea Military Academy
Korea

1. Introduction

In mobile robotics, obstacle avoidance problem has been studied as a classical issue. In practical applications, a mobile robot should move to a goal in the environment where obstacles coexist. It is necessary for a mobile robot to safely arrive at its goal without being collided with the obstacles. A variety of obstacle avoidance algorithms have been developed for a long time. These algorithms have tried to resolve the collisions with different kinds of obstacles. This chapter classifies the types of obstacles and presents characteristics of each type. We focus on the new obstacle type that has been less studied before, and apply representative avoidance strategies to solve this new type of obstacle avoidance problem. The results of different strategies would be compared and evaluated in this chapter.

2. Classification of obstacles

This section summarizes previous points of view on obstacles, and presents a new angle on obstacles for advanced robot navigation. A robotic system often suffers from severe damage as result of a physical collision with obstacles. Thus, there is no doubt that robots should have a perfect competency to avoid all kinds of obstacles. In robotics, an obstacle is generally defined as a physical object that is in the way or that makes it hard for robot to move freely in a space, and thereby classified into *static and moving obstacles*: the former are sometimes subdivided into stationary (or fixed) and movable ones of which position can be changed by force from a robot. Also, moving robots are often separated from moving obstacles, since their motion can be changed by themselves.

On the basis of this definition, various strategies have been presented to resolve anticipated physical collisions systematically. Since the majority of strategies focus on generating immediate reaction to environments, obstacles are considered static for an instant although they are moving in most previous studies. Recently, some studies formalized way to explicitly consider current velocity of moving obstacles for coping with real collisions caused by the static assumption. This chapter views obstacles from a different standpoint and introduces a concept of *non-rigid obstacles*, which differs from the conformable obstacle in that the concept covers various physical entities as well as the shape (or boundary) of obstacle. Fig. 1 shows these views on obstacles and details are discussed below.

2.1 Rigid obstacles

Early studies on robot motion assumed that a robot is the only moving object in the workspace and all obstacles are fixed and distributed in a workspace (Latombe, 1991). Under the

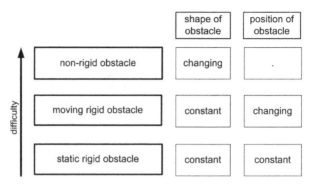

Fig. 1. Classification of obstacles

assumption, obstacle avoidance is automatically solved if we solve the global path planning problem in which the robot is assumed to have complete knowledge of the environment and generates a full path from its start to goal. The famous examples include the following approaches: probabilistic road map (Kavraki et al., 1996), cell decomposition (Boissonnat & Yvinec, 1998), grid cells (Hee & Jr., 2009), and rapidly-exploring randomizing tree (LaValle, 1998). To realize the approaches, several systematic tools for path generation were developed: visibility graph, Voronoi diagram, random sampling method, gradient method, node-edge graph, and mathematical programming (Ge & McCarthy, 1990). The final objective of the approaches is to find the global optimal (or shortest) one in the path candidates given by an optimization algorithm, such as Dijkstra and A* method.

Some studies relaxed the assumption of complete knowledge about environments to take into consideration the case that there is no device to provide the complete knowledge. Accordingly, the studies focused on following a detour (or local path) to avoid a detected static obstacle, instead of finding the shortest path to its goal. As a result, the obstacle avoidance is separated from the global path planning, which implies that it can be performed by on-line planning. The simplest strategy for the static obstacle avoidance with a limited sensing range may be the bug algorithm in which the robot is controlled to simply follow the contour of an obstacle in the robot's way (Lumelsky & Skewis, 1990). As with the bug algorithm, other avoidance strategies involving such obstacles create an immediate reaction to nearby environments for robot navigation. The popular methods include vector field histogram (Borenstein & Koren, 1991), nearness diagram (Minguez & Montano, 2004), fuzzy rules (Lee & Wang, 1994), force (or potential) fields (Wang et al., 2006), and dynamic window (Fox et al., 1997). In fact, these tools have been widely used for even the following moving obstacle avoidance problem by virtue of their simplicity in formulation and low operational speed of robots which permits the assumption that moving obstacles can be seen as static ones for a short time.

Moving obstacle avoidance differs from the static case in that an irreversible dimension - time - should be added into problem formulation. Like the static obstacle avoidance problem, the solution for the moving case can be computed globally or locally according to the assumption of the complete knowledge of environments. The configuration-time (CT) space presented in fig. 2 and 3 is a well-known tool for global problem solving. The CT space is a three-dimensional space in which a time axis is added into a xy-plane where a robot is moving. Static obstacles are represented as a straight pillar as shown in fig. 2 when they are represented with respect to CT space. On the other hands, if a moving obstacle is modeled as a circle, its representation with respect to CT space is an oblique cylinder as shown in fig. 3,

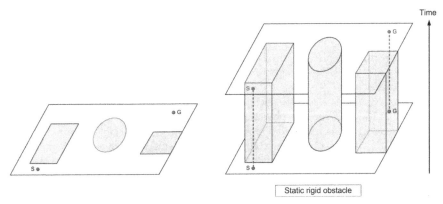

Fig. 2. A configuration-time space of static rigid obstacles

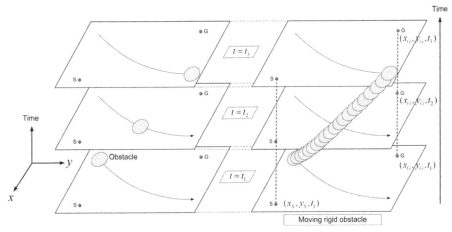

Fig. 3. A configuration-time space of moving rigid obstacles

because the position of the obstacle is varied with time along its path. These representations indicate that the problem of global moving obstacle avoidance is reduced to finding a 3-D geometric path obeying two conditions: safety and time efficiency. Sampling, combinatorial, and velocity-tuning methods have been introduced to solve the geometric problem (Lavalle, 2006).

Local moving obstacle avoidance is split into two groups, according to whether or not obstacles can be seen static for a short time, as mentioned above. Several recent papers argued that the problem can be made more realistic by explicitly considering the robot's velocity and acceleration, and included the constraints to problem formulation. The collision representation tools based on the consideration include the velocity obstacle (VO) (Fiorini & Shiller, 1993), inevitable collision states (ICS) (Martinez-Gomez & Fraichard, 2009) and the triangular collision object (TCO) (Choi et al., 2010). The major difference is that object shapes representing potential collisions are determined by obstacle's velocity, unlike the tools for the static obstacle avoidance problem. In the benchmarking (Martinez-Gomez & Fraichard, 2009), it was shown that the method using the VO is superior to the methods for static obstacle avoidance because it takes into account objects' future behavior, and suggested that the future motion of moving obstacles should explicitly been deal with. So far, we have briefly reviewed conventional rigid obstacle types and avoidance strategies which are popularly used in robotics. Now turn our attention to a new type, non-rigid obstacle in the following section.

2.2 Non-rigid obstacles

In this section, we explain a new obstacle type, non-rigid obstacle. This concept emerges as a consequence of the need for non-physical collisions, such as observation. More specifically, robots only need to have considered a single physical element, obstacle's boundary, for safe navigation to date, but they now have to consider other elements, such as agent's visibility, for smart navigation. Thus, to accommodate such a need, we here define a new term, *non-rigid obstacle*, which is an abstract object of which boundary is changeable and determined by an element involving an obstacle.

We explain this concept, non-rigid obstacle, through a moving observer with omnidirectional view. In a infiltration mission, the robot should covertly trespass on the environment without being detected by sentries. As a reverse example, in a security mission, the most important thing is that robots are not observed (or detected) by the intruder until it reaches the intruder (Park et al., 2010). In this way, the robot can covertly capture or besiege the intruder who would try to run away from the robot when detecting the approaching robot. It causes that robots can closely approach the intruder without being detected, and the intruder cannot recognize himself being besieged until the robots construct a perfect siege. In these cases, the robot should avoid the intruder's field of view during the navigation, rather than its physical boundary, which is called *stealth navigation*. This signifies that the proposed concept of non-rigid obstacle could be exceptionally helpful in planning an advanced robot motion.

The visible area determined by the field is a changeable element, especially in a cluttered indoor environment. So, we can regard the area as an abstract non-rigid obstacle that robots should avoid. This kind of obstacle can be generated by any element the obstacle has, which is a remarkable extension of the conventional concept on obstacles. Fig. 4 illustrates an example of the changes of visible areas with time traveled (or z-axis) in a CT space. As shown in the figure, the shape (or boundary) of visible area is greatly varied with the observer's motion which is a set of position parameterized by time. For this reason, it is needed to develop a new strategy to avoid the highly changeable obstacle, which will be discussed in detail in Section 3. If the robot succeeds in the avoidance with a plausible strategy, it gives a completely new value to a robotic system.

3. Avoidance strategies for non-rigid obstacle avoidance

As we mentioned in section 2, researches on obstacle avoidance problem have been focused on dealing with rigid (static or moving) obstacles. On the other hand, researches on avoidance strategies of non-rigid obstacles scarcely exist except grid-based methods (Marzouqi & Jarvis, 2006; Teng et al., 1993) a few decades ago. These methods were based on the grid map which approximates the environment to spatially sampled space. In this grid map, all possible movements of the robot were taken into consideration to find the optimal(fastest) motion of the robot to the goal. This brought about high burden of computation time. In addition, the arrival time of the planned motion and the computation time highly depended on the grid size.

Since then, any other strategies have not been developed for non-rigid obstacle avoidance. Even though other methods have not been suggested, we can apply rigid obstacle avoidance

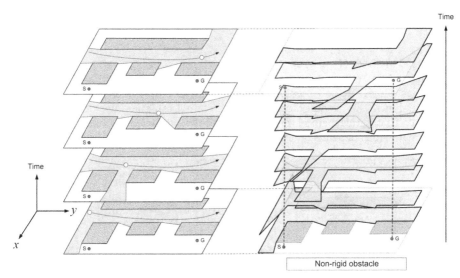

Fig. 4. A configuration-time space of non-rigid obstacles

algorithms to the problem of non-rigid obstacle avoidance. On a big scale, these algorithms can be classified into four categories: reactive method, grid-based method, randomization method, and path-velocity decomposed method. We will explain this classification of algorithms and how they can be applied to non-rigid obstacle avoidance.

3.1 Reactive method

Reactive methods utilize nearby environment information for robot navigation, i.e., the robot has a sensing capability with a limited coverage. In the case where the robot is sent to a unknown region for exploration or reconnaissance, it is reasonable to assume that information of the environment is not known beforehand. In addition, if the robot has no communication connection with other sensors on the environment, then it has to rely on the nearby information acquired from its own sensor. In this case, the robot should find a suitable motion by making full use of given information in order to attain two purposes: avoiding obstacles and moving to its goal. In other words, it can be widely used when the information of the environment is limited and not deterministic, whereas other methods, which will be explained in following sections, are based on the assumption that the information of environment is known in advance.

This reactive method can also be called short-term planning because it plans a few next steps of robot motion. The planner focuses on creation of immediate respondence of the robot in order to avoid obstacles and move to the goal. On a rough view, the reactive method yields instantaneous moving direction by synthesizing attractive attribute to the goal and repulsive attribute from obstacles. In non-rigid obstacle avoidance, the robot is planned to be repulsive from non-rigid obstacles. As mentioned in section 2.1, there have been many researches to find reactive respondence of the robot, such as vector field histogram(VFH) (Borenstein & Koren, 1991), force field (Wang et al., 2006), dynamic window (Fox et al., 1997), and behavior-based robotics (Arkin, 1985). In this chapter, we implemented one of representative reactive methods, vector field histogram, for comparison. It constructs a polar histogram, which contains nearby environment information, and finds suitable detour direction towards its goal.

3.2 Grid-based method: (Teng et al., 1993)

Grid-based methods use spatially sampled environment map as mentioned earlier. This algorithm has a merit that all possible robot motion can be considered in the grid map at the expense of increasing computation time. Therefore it can guarantee optimal solution in the grid map. That is, this method can have the completeness even though it is limited to the given grid map which depends on the shape and the size of grids.

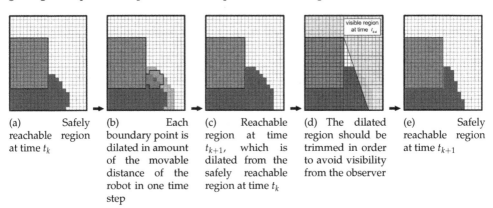

| (a)   Safely reachable region at time t_k | (b)   Each boundary point is dilated in amount of the movable distance of the robot in one time step | (c)   Reachable region at time t_{k+1}, which is dilated from the safely reachable region at time t_k | (d) The dilated region should be trimmed in order to avoid visibility from the observer | (e)   Safely reachable region at time t_{k+1} |

Fig. 5. Construction procedure of safely reachable region(purple) in grid-based method

Differently from other methods, this grid-based algorithm was designated for solving this non-rigid obstacle avoidance and introduced in (Teng et al., 1993). In this algorithm, the concept of the *safely reachable region* was introduced and used. It represents the region that the robot is able to arrive without being detected by the observer at each time step. The planner sequentially constructs safely reachable region of each time step until the last safely reachable region contains the final goal. Fig. 5 represents the procedure for obtaining safely reachable region from a previous one. Let the purple region of fig. 5(a) be the safely reachable region at time t_k. This region is dilated to the reachable region from it one time step later, which is shown in fig. 5(c). This dilated region is trimmed by the visible area of the observer at time t_{k+1}. Then it becomes the region that the robot can arrive without being detected by the observer at time t_{k+1}, which is shown in fig. 5(e). In this way, all possible safely reachable grid points were calculated at each time, and the shortest time for arrival is yielded.

3.3 Randomization method: modified RRT

In order to avoid a high computational burden of brute-force technique, randomization methods have been developed in robot motion planning. In these methods, free configurations are randomly selected and discriminated whether they can be via configurations that the robot is able to safely pass by. These kinds of algorithms have been developed for a long time, such as probabilistic road map(RPM) (Kavraki et al., 1996) and rapidly-exploring randomizing tree(RRT) (LaValle, 1998). These randomization methods could be usefully adopted for solving non-rigid obstacle avoidance problem in order to prevent burgeoning computational burden. In this chapter, we have modified RRT algorithm in order to cope with non-rigid obstacles which change its shape while time passes. The modified RRT algorithm for solving non-rigid obstacle avoidance problem is described in table 1.

```
0:   s_init ← (t_init, q_init);
1:   G.init(s_init);
2:   while
3:       q_rand ← RAND_FREE_CONF();
4:       s_near = (t_near, q_near) ← NEAREST_VERTEX(q_rand, G);
5:       v_rand ← RAND_VEL();
6:       if CHECK_SAFETY(s_near, q_rand, v_rand) then
7:           t_new ← NEW_TIME(s_near, q_rand, v_rand);
8:           s_new ← (t_new, q_rand);
9:           G.add_vertex(s_new);
10:          G.add_edge(s_near, s_new);
11:          v_g_rand ← RAND_VEL();
12:          if CHECK_SAFETY(s_new, q_goal, v_g_rand) then
13:              t_arrival ← NEW_TIME(s_new, q_goal, v_g_rand);
14:              s_goal ← (t_arrival, q_goal);
15:              G.add_vertex(s_goal);
16:              G.add_edge(s_new, s_goal);
17:              break;
18:          end if;
19:      end if;
20: end while;
```

Table 1. Construction procedure of rapidly-exploring randomizing tree G for non-rigid obstacle avoidance

In the algorithm, the tree G is constructed with randomly selected vertexes. Each vertex s consists of time t and configuration q, i.e., $s = (t, q)$. In non-rigid obstacle avoidance problem, the time t when the robot passes by the configuration q should be coupled with q as a vertex. Thus, the time t also should be selected randomly. In this algorithm, t is calculated as the time at which the robot is able to arrive at q with the randomly selected velocity. Each randomly selected vertex becomes component of the tree G, and the tree G is continuously extended until new selected vertex is able to be linked with the goal vertex. After construction of tree G is completed, the connection from the start vertex to goal vertex is determined as the robot's movement.

In table 1, RAND_FREE_CONF() returns randomly selected free configuration. NEAREST_VERTEX(q,G) returns the vertex $s_r = (t_r, q_r)$ in G, which has the minimum distance from q to q_r. RAND_VEL() is the function that yield randomly selected velocity v which should satisfy the robot's physical constraints, i.e., $0 \leq v \leq v_{max}$. In this procedure, the probability distribution function $f_V(v)$ for selecting v was designed as follows because it is beneficial for the robot to move as fast as possible.

$$f_V(v) = \frac{2v}{v_{max}^2}, \quad 0 \leq v \leq v_{max}$$

CHECK_SAFETY($\mathbf{s}_1$,$\mathbf{q}_2$,v) returns whether it is possible that the robot is able to safely move from the configuration $\mathbf{q}_1$ of vertex $\mathbf{s}_1$ to $\mathbf{q}_2$ with the velocity v or not. NEW_TIME($\mathbf{s}_1$,$\mathbf{q}_2$,v) returns the time t_r taken for the robot to move from $\mathbf{q}_1$ of $\mathbf{s}_1$ to $\mathbf{q}_2$ with the speed v. In this chapter, the configuration $\mathbf{q}$ of the robot is two dimensional point $\mathbf{p}$ because the robot is assumed to be holonomic and has circular shape. In this case, returned time t_r is calculated as follows.

$$t_r = t_1 + \frac{\| \mathbf{p}_2 - \mathbf{p}_1 \|}{v}$$

3.4 Path-velocity decomposed method

Lastly, we can consider the path-velocity decomposed(PVD) method. This concept was firstly introduced in (Lee & Lee, 1987) and has been developed for resolving collisions of multiple robots in (Akella & Hurchinson, 2002; Lavalle & Hurchinson, 1998). This method decomposes the robot motion into the path, which is a geometric specification of a curve in the configuration space, and the velocity profile, which is a series of velocities according to the time following the path. In other words, the PVD method firstly plans the path, and adjusts the velocity of the robot following this path. It reduces computation burden by not guaranteeing the safety condition when planning the path at the cost of optimality. This safety condition is considered only in planning of velocity profile. If we predefine the geometric road map (topological graph) on the environment, we can easily plan the robot's path on the road map and find the velocity profile that guarantees safely movement of the robot following the path. This method used for solving moving rigid obstacle avoidance also can be applied to non-rigid obstacle avoidance problem without difficulty.

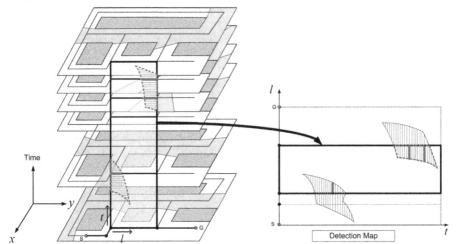

Fig. 6. Construction of the detection map

(Park et al., 2010) has proposed the concept of detection map. The detection map is a two-dimensional(time and point on the path) map that represents whether each time and point on the path is detected by the observer or not. It depends on the path chosen on the road map. Fig. 6 shows the construction of the detection map on the predetermined path. When the detected intervals on the path are accumulated according to the time, the detection map is constructed. In the detection map, the shaped regions represents the time and points on the path that are detected by the observer, which is called detection region. We can draw the robot's trajectory on the detection map without intersecting these detection regions in order to make the robot safely move to its goal. After the detection map is constructed, the safely velocity profile can be found in a shorter time because moving direction is restricted to one dimension.

The planning method of the velocity profile can be selected in a various way. (Park et al., 2010) applied the grid-based method to this velocity profile planning. The concept in (Teng et al., 1993), which constructs safely reachable region by dilation and trimming in order to find the optimal solution on the grid map, is adopted to solve the similar problem with the one-dimensional moving direction. The points on path is sampled with uniform interval and examined whether it is safely reachable by the robot for each discretized time step. This grid-based method also finds the velocity profile that has the shortest arrival time. We note that this optimality is valid only when the predetermined path could not be modified by the planner.

The randomization method could be proposed in velocity profile planning. Without considering all possible safe movement of the robot, we can find the robot's safe trajectory by randomly selecting via points on the detection map. In this chapter, we applied the modified RRT method explained in section 3.3 t velocity profile planning. In this application, the one-dimensional value is randomly selected as via point that the robot would pass. The RRT would be constructed on the detection map without intersecting detection regions. The detection map makes it easy to discriminate whether tree link is safe or not.

The method of selecting path on the road map also could be suitably adopted. (Park et al., 2010) has selected the path that has the shortest moving distance. If the safe velocity profile did not exist on the selected path, the next shortest path was examined in consecutive order. However, it often led to a situation that not a few path had to be examined because this method completely ignored the safety condition on the path planning step. We proposed the method to select the path highly possible to have safe trajectory of the robot. We designed the cost of each link on the road map as follows. Let $E_{i,j}$ be the link between node i and node j on the road map.

$$cost(E_{i,j}) = length(E_{i,j}) * (detection\_ratio(E_{i,j}))$$

$$\text{where } detection\_ratio(E_{i,j}) = \frac{area\ of\ detection\ region\ on\ the\ detection\ map\ of\ E_{i,j}}{area\ of\ whole\ detection\ map\ of\ E_{i,j}}$$

Compared to the method which considered only path distance, this cost function makes the path planner to consider not only distance but also safely condition. If we select the path that has the least cost on the road map, this path would be relatively less detected by the observer. If a safe velocity profile is not found on the selected path, then the other path which has next least cost could be found and examined again.

4. Simulation results

We have conducted a computer simulation on the environment in fig. 7. In this simulation, the visible area of the observer is considered as a example of the non-rigid obstacle. The robot is planned to move stealthy from the observer. The observer is patrolling on the environment following a specific path with the speed of 2.0m/s. It is assumed that the robot has a maximum speed of 4.0m/s and it can change its speed and its moving direction instantaneously. We assumed that the information of the environment and the non-rigid obstacle is known to the

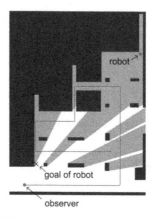

Fig. 7. Simulation environment (27.5m x 36.5m) and the patrol path of the observer

robot. This computer simulation is conducted on a commercial computer with 3.34GHz CPU and 3.00GB RAM.

Table 2 shows the simulation results according to algorithms from reactive method to path-velocity decomposed method. In table 2, the *exposure rate* means a ratio of the time interval that the robot is detected by the observer to the whole arrival time. In addition, we define *safety* as the distance between the robot and the point which is contained in the visible area of the observer and is the closest to the robot. When the robot is contained in the visible area of the observer, i.e., the robot is detected by the observer, the safety value is set to zero. In table 2, the average value of safety during the operation time is revealed. Also, the computation times of each method are shown in table in order to analyze practical implementation feasibility of each method.

| | Reactive | Grid-based | Randomization | Path-velocity decomposed | |
| | | | | Grid-based | Randomization |
	(VFH)	(Teng et al., 1993)	(Modified RRT)	(Park et al., 2010)	(Modified RRT)
Arrival time	10.8s	10.4s	20.1s	16.2s	16.6s
Exposure rate	3.0%	0.0%	0.0%	0.0%	0.0%
Average safety	2.97m	2.88m	3.14m	3.32m	3.40m
Computation time	0.01s	33.67s	17.15s	path planning: 3.70s	
				0.01s	0.01s

Table 2. Simulation results

In table 2, the exposure rate had non-zero value only when the reactive method was applied. This is because the reactive method does not utilize whole information of visible areas of the observer. It assumes that the robot is able to sense only nearby information of environment and visible area. In this simulation, the robot's sensing coverage is assumed to be omnidirectional with 3.0m radius. Therefore, it is possible that the robot is placed in the situation that it is too late for avoiding the visible area that emerges suddenly near by. At the cost of this dangerous situation, the reactive method does require a negligible amount of the computation time. More detailed results are shown in fig. 8 and fig. 9. As we can see in fig. 8(b), and 8(c), the robot's speed and direction were changed extremely for a whole operation time. In fig. 9, the robot detects the non-rigid obstacle in the rear of itself at time 5.7(s), nevertheless, it could not avoid the obstacle due to the lack of speed, and it collided with the obstacle at time 6.0(s). We note that this chapter assumes that non-rigid obstacle is also detected by the sensor of the robot in order to apply the reactive method for rigid obstacle avoidance to non-rigid obstacle avoidance problem. In real application, some non-rigid obstacle such as observer's field of view could not be detected by the sensor. As this result shows, the robot has a difficulty in stealth navigation when it has a limited sensing capability of the environment.

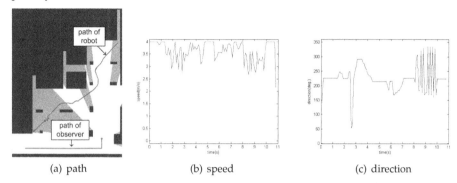

(a) path   (b) speed   (c) direction

Fig. 8. The reactive method (VFH) results

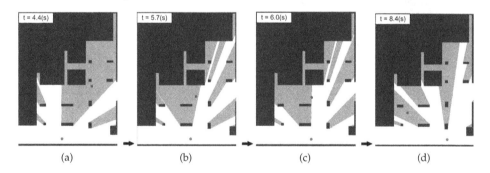

(a)   (b)   (c)   (d)

Fig. 9. The movement of observer(red) and robot(blue) in the reactive method result

From the results in table 2, we can find that the arrival time was the least when the grid-based method was applied. This is because the grid-based method considers all possible motions of the robot and finds the fastest motion among them, i.e., it guarantees the optimality and completeness. However, these are valid only on given grid map. That is, if the grid map is set in a different way, the robot's fastest motion and the arrival time would become different. Since it examines all grid cells whether they are possibly reached safely by the robot at each time, it needs high computational burden. In table 2, the computation time of the grid-based method is higher than that of any other methods. More detailed results are shown in fig. 10. The speed and direction of the robot were not changed extremely. Between time 6.8(s) and 9.1(s), the robot remained stationary in the shaded region until the obstacle was away from itself. The path in fig. 10(a) shows that the robot was able to almost directly traverse the environment to the goal without being detected by the observer. The performance in path length and arrival time is the best in the grid-based method. Fig. 11 shows safely reachable regions changing in accordance with the passed time.

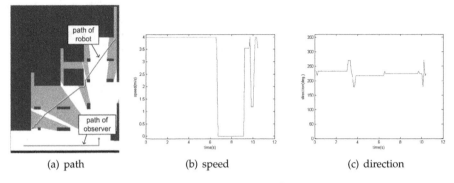

(a) path   (b) speed   (c) direction

Fig. 10. The grid-based method results(grid size is $64cm^2$/cell)

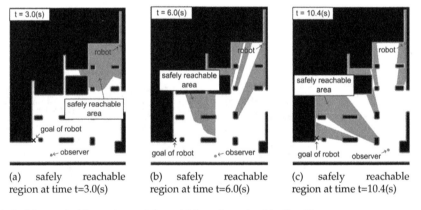

(a) safely reachable region at time t=3.0(s)

(b) safely reachable region at time t=6.0(s)

(c) safely reachable region at time t=10.4(s)

Fig. 11. Safely reachable regions of the grid-based method in fig. 10

The result of randomization method in table 2 has the lowest performance in the arrival time. This result is not deterministic and can be changed in each trial. There is possibility that relatively better performance can be yielded, but it could not be guaranteed. This method does not purpose on greatly reducing the arrival time, but it concerns a practical computation time. In table 2, modified RRT algorithm needed not a low computation time. This is because it needs relatively much time to check that the tree link candidate is safe for the movement of the robot. In order to reduce the computation time in the randomized method, it is necessary to develop a specialized randomization method to solve the non-rigid obstacle avoidance problem. Fig. 12 shows the result of modified RRT method. In fig. 12(a) and 12(c), the path of the robot was planned as zigzagged shape because via points are selected randomly. As shown in fig. 12(b), the segments between via points have respectively uniform speeds which are also selected randomly.

Lastly, we implemented the PVD method in the same environment. The road map of the environment was predefined, and the path with the least cost on the road map was selected as shown in fig. 13(a). We can see that the path with the least cost is selected as series of

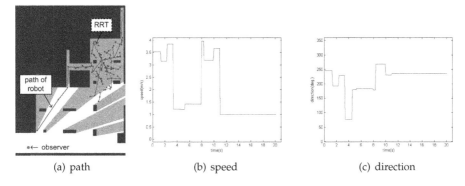

(a) path   (b) speed   (c) direction

Fig. 12. The randomized method (modified RRT) results

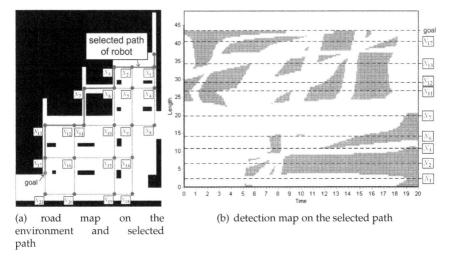

(a) road map on the environment and selected path

(b) detection map on the selected path

Fig. 13. The road map and the detection map

edges that are in the vicinity of walls, because they have relatively small detection ratios. The detection map corresponding to the selected path was constructed as shown in fig. 13(b). In the detection map, the colored regions, which are detection regions, are the set of time and point that is detected by the observer on the determined path. The trajectory of the robot following this path could be planned on the detection map without traversing detection regions.

As shown in table 2, the arrival time of the grid-based result in PVD method was larger than that of the grid-based method. This is because the PVD method restricts the path to be selected on the predefined roadmap, not among whole possible movement. Therefore, the PVD method has the lower performance than the grid-based method even though it also finds the optimal solution on the determined path. However, its computation time is much smaller than the grid-based method because it reduces the dimension of the grid map. This planned result is shown in fig. 14(a). The movement of the observer and the robot in this result is shown in fig. 15. In fig. 15(b) and 15(c), we can see that the robot remains stationary because

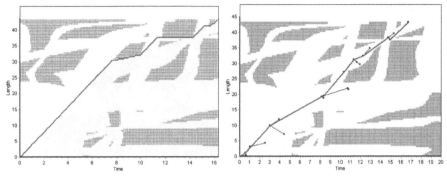

(a) planned trajectory of the grid-based method (b) planned trajectory of the randomization
method

Fig. 14. The trajectory results of two detail velocity profile planning algorithm in the
path-velocity decomposed(PVD) method

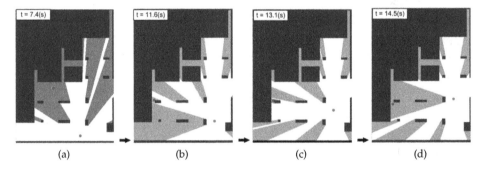

Fig. 15. The movement of observer(red) and robot(blue) in the grid-based result of
path-velocity decomposed method

the observer's field of view obstructs the front and the rear of the robot. In fig. 15(d), the robot
started to move after the observer's field of view is away from itself.

In the case of the randomization result in PVD method in table 2, it not only needs much lower
computation time but also yields the better performance than the modified RRT methods. The
randomization methods cannot guarantee the optimality, and its performance is on a case by
case basis. Therefore, it is possible that the case restricting path in PVD method has the shorter
arrival time than the case having whole opportunity of path selection. This planned result is
shown in fig. 14(b).

5. Conclusion

In this chapter, we classified the type of obstacles into three categories: static rigid obstacles,
moving rigid obstacles, and non-rigid obstacles. However, most researches of obstacle
avoidance have tried to solve the rigid obstacles. In reality, the non-rigid obstacle avoidance
problem is often emerged and needed to be solved, such as stealth navigation. We adopted
the representative algorithms of classic rigid obstacle avoidance, and applied to non-rigid

obstacle avoidance problem. We examined how the inherent characteristics of each algorithms are revealed in non-rigid obstacle avoidance and evaluated the performance. The result in the reactive method informed us that limited information of the environment and obstacle greatly hinders the robot from avoiding non-rigid obstacles. The grid-based method could find the optimal safe motion of the robot at the cost of high computation burden. The randomization method gave priority to reducing computational burden over improving the performance. Finally, the path-velocity decomposed(PVD) method greatly reduced computation burden by abandoning the diversity of path selection. This method does not cause great loss in arrival time. This method makes the robot avoid the obstacle by adjusting its speed. In velocity profile planning, several detail methods, such as the grid-based method and the randomization method, can be applied with practical computation time. When complete information of the environment and the obstacle is given, this PVD method could be efficiently used.

6. References

Akella, S. & Hurchinson, S. (2002). Coordinating the motions of multiple robots with specified trajectories, *Proceedings of IEEE International Conference of Robotics and Automation*, New Orleans, LA, pp. 3337–3344.

Arkin, R. C. (1985). *Behavior-based Robotics*, MIT Press.

Boissonnat, J. D. & Yvinec, M. (1998). *Algorithm Geometry*, Cambridge University Press.

Borenstein, J. & Koren, Y. (1991). The vector field histogram - fast obstacle avoidance for mobile robot, *Journal of Robotics and Automation* Vol. 7(No. 3): 278–288.

Choi, J. S., Eoh, G., Kim, J., Yoon, Y., Park, J. & Lee, B. H. (2010). Analytic collision anticipation technology considering agents' future behavior, *Proceedings of IEEE International Conference on Robotics and Systems*, pp. 1656–1661.

Fiorini, P. & Shiller, Z. (1993). Motion planning in dynamic environments using the relative velocity paradigm, *Proceedings of IEEE International Conference on Robotics and Automation*, Vol. Vol. 1, pp. 560–566.

Fox, D., Burgard, W. & Thrun, S. (1997). The dynamic window approach to collision avoidance, *IEEE Robotics and Automation Magazine* Vol. 4(No. 1): 23–33.

Ge, Q. J. & McCarthy, J. M. (1990). An algebraic formulation of configuration-space obstacles for special robots, *Proceedings of International Conference on Robotics and Automation*, pp. 1542–1547.

Hee, L. & Jr., M. H. A. (2009). Mobile robots navigation, mapping, and localization part i, *Encyclopedia of Artificial Intelligence* pp. 1072–1079.

Kavraki, L. E., Svestka, P., Latombe, J. C. & Overmars, M. H. (1996). Probabilistic roadmaps for path planning in high-dimensional configuration space, *IEEE Transactions on Robotics and Automation* Vol. 12(No. 4): 566–580.

Latombe, J. C. (1991). *Robot Motion Planning*, Kluwer academic publishers.

LaValle, S. M. (1998). Rapidly-exploring random trees: A new tool for path planning, *Technical report 98-11, Computer Science Department, Iowa State University* .

Lavalle, S. M. (2006). *Planning Algorithms*, Cambridge University Press.

Lavalle, S. M. & Hurchinson, S. A. (1998). Optimal motion planning for multiple robots having independent goals, *IEEE Transactions on Robotics and Automation* Vol. 14(No. 6): 912–925.

Lee, B. H. & Lee, C. S. G. (1987). Collision-free motion planning of two robots, *IEEE Transactions on Systems Man and Cybernetics* Vol. 17(No. 1): 21–31.

Lee, P. S. & Wang, L. L. (1994). Collision avoidance by fuzzy logic for agv navigation, *Journal of Robotic Systems* Vol. 11(No. 8): 743–760.

Lumelsky, V. & Skewis, T. (1990). Incorporating range sensing in the robot navigation function, *IEEE Transactions on Systems Man and Cybernetics* Vol. 20: 1058–1068.

Martinez-Gomez, L. & Fraichard, T. (2009). Collision avoidance in dynamic environments: an ics-based solution and its comparative evaluation, *Proceedings of IEEE International Conference on Robotics and Automation*, Kobe, pp. 100–105.

Marzouqi, M. S. & Jarvis, R. A. (2006). New visibility-based path-planning approach for covert robotics navigation, *IEEE Transactions on Systems Man and Cybernetics* Vol. 24(No. 6): 759–773.

Minguez, J. & Montano, L. (2004). Nearness diagram navigation: collision avoidance in troublesome scenarios, *IEEE Transactions on Robots and Automation* Vol. 20(No. 1): 45–59.

Park, J., Choi, J. S., Kim, J., Ji, S.-H. & Lee, B. H. (2010). Long-term stealth navigation in a security zone where the movement of the invader is monitored, *International Journal of Control, Automation and Systems* Vol. 8(No. 3): 604–614.

Teng, Y. A., Dementhon, D. & Davis, L. S. (1993). Stealth terrain navigation, *IEEE Transactions on Systems Man and Cybernetics* Vol. 23(No. 1): 96–113.

Wang, D., Liu, D. & Dissanayake, G. (2006). A variable speed force field method for multi-robot collaboration, *Proceedings of IEEE International Conference on Intelligent Robots and Systems*, pp. 2697–2702.

Part 2

Path Planning and Motion Planning

7

Path Planning of Mobile Robot in Relative Velocity Coordinates

Yang Chen[1,2,3], Jianda Han[1] and Liying Yang[1,3]
[1]State Key Laboratory of Robotics, Shenyang Institute of Automation,
Chinese Academy of Sciences, Shenyang
[2]Department of Information Science and Engineering,
Wuhan University of Science and Technology, Wuhan
[3]Graduate School, Chinese Academy of Sciences, Beijing
China

1. Introduction

Path planning of mobile robot in dynamic environment is one of the most challenging issues. To be more specific, path planning in multi-obstacle avoidance environment is defined as: given a vehicle A and a target G that are moving, planning a trajectory that will allow the vehicle to catch the target satisfy some specified constrains while avoiding obstacle O, and each of the obstacles can be either mobile or immobile in the environment. The corresponding problem is named target-pursuit and obstacles-avoidance (TPOA) and will be researched extensively in this chapter.

The traditional method, such as probability road map, can achieve a successful path in 2D static environments. The planning process using this method generally consists of two phases: a construction and a query phase. In construction stage, the workspace of the robot is sampled randomly for generating candidate waypoints. In the query stage, the waypoints between the start and goal position are connected to be a graph, and the path is obtained by some searching algorithm, such as Dijkstra, A* algorithm and so on. Hraba researched the 3D application of probability road map where A* algorithm is used to find the near-optimal path (Hrabar, 2006). Although probability road map method is provably probabilistically complete (Ladd & Kavraki, 2004), it does not deal with the environment where the information is time-varying. The underlying reason is that this method only focuses on the certain environment. Once some uncertainty appears in the robot workspace, probability road map can not update with the changing environment and plan a valid trajectory for the mobile robot, never an optimal path.

Artificial potential field is another traditional method which is generally used in both 2D and 3D environment. The mechanism that the robot is driven by attractive and repulsive force in a cooperative way is simple and often works efficiently even in dynamic environment. Kitamura et al. construct the path planning model based on the artificial potential field in three-dimensional space which is described by octree (Kitamura et al, 1995). Traditionally, artificial potential field applies in two dimensions extensively. Also some other field concepts are invented. For example, there are harmonic potential functions (Kim & Khosla, 1992; Fahimi et al, 2009; Cocaud et al, 2008; Zhang & Valavanis, 1997), hydrodynamics (Liu et al, 2007), gradient field (Konolige, 2000), and virtual force field (Oh et al., 2007). Unfortunately, path planning approach based on the function family of potential field cannot obtain the optimal objective function which, in turn, cannot guarantee the desired optimal trajectories. Additionally, some of them, for example, potential panel method and harmonic potential function, can plan a path for an autonomous vehicle, but the computing burdens is huge and real time performance hardly satisfies the practical requirement.

Inspired by biological intelligence, many approaches, such as ant colony optimization (ACO) (Chen et al., 2008), particle swarm optimization (PSO) (Jung et al., 2006), genetic algorithm (GA) (Allaire et al., 2009), evolution algorithm (EA) (Zhao & Murthy, 2007), and their combination, are introduced to solve the path planning problem. They mostly rely on the stochastic searching, known as non-deterministic algorithm. These methods will eventually find an optimal solution, but no estimate on the time of convergence can be given. Thus, it may take long even infinite time to find the best solution. Furthermore, all of them have a great number of parameters to tune and that is never an easy job particularly when users are short of prior knowledge. Some comparisons (Allaire et al., 2009; Krenzke, 2006) show that the expensive calculations limit their real application.

Another kind of method is mathematic programming. Based on the relative velocity coordinates (RVCs), linear programming (LP) and mixed integer linear programming (MILP) can be employed for path planning problem. For example, Wang et al. converted the 2D path planning of an unmanned underwater vehicle to constrained optimization or semi-infinite constrained optimization problem (Wang et al., 2000). Zu et al. discussed the path planning in 2D and an LP method was proposed for the problem of dynamic target pursuit and obstacle avoidance (Zu et al., 2006). This method tried to plan the variables of the linear acceleration and the angular acceleration of a ground vehicle. Schouwenaars et al. proposed a MILP formulation with receding horizon strategy where a minimum velocity and a limited turn rate of aircraft are constrained (Schouwenaars et al., 2004). However, these results still focused on the two dimensions.

In this chapter, we consider the same problem of TPOA but in three dimensions. To our problem, the uncertain environment has one target or some of them, denoted by G, and many obstacles O that are all velocity-changeable with their moving actions. The aim of the vehicle A is to find an optimal path for pursuing the target while, at the same time, avoiding collision threaten from obstacles. The position and velocity of movers, including the target and obstacles, are assumed known or estimated at current time. To be more specific, it is assumed that the noise contained in the data can be eliminated with certain filters. However, this paper has no intend to discuss the filtering algorithm for achieving the feasible data of on-board sensors. The trajectory from a start to a destination location typically needs to be computed gradually over time in the fashion of receding horizon (Schouwenaars et al., 2004) in which a new waypoint of the total path is computed at each time step by solving a linear programming problem. The target and obstacles, probably static or dynamic, are modeled by spheres having a certain radius for their impact area, and the vehicle is modeled by a mass point. In order to optimize the vehicle's acceleration online, we construct a linear programming model in Cartesian orthogonal coordinates based on relative velocity space (Fiorini & Shiller, 1998).

This chapter is organized as follows. First, the relative velocity coordinates are introduced in both two dimensions and three dimensions and then the TPOA principles are introduced, including obstacle-avoidance and target-pursuit principles. The formulation of the standard linear programming is mathematically introduced in Section 3.1, and the path planning model is described in detail in Section 3.2. Simulations of path planning in 2D and 3D are

A, G, O	Vehicle, target, obstacle
τ	Planning period
k	Time step
N	Sum of obstacles
$\mathbf{V}_A$	Vehicle velocity
$V_A = \|\mathbf{V}_A\|$	Magnitude of the vehicle velocity
$\Delta\mathbf{V}_A$	Vehicle acceleration
$\mathbf{V}_O$	Obstacle velocity
$V_O = \|\mathbf{V}_O\|$	Magnitude of the obstacle velocity
$\mathbf{V}_G$	Target velocity
$V_G = \|\mathbf{V}_G\|$	Magnitude of the target velocity
$\mathbf{V}_{AO} = \mathbf{V}_A - \mathbf{V}_O$	Vehicle's relative velocity to the obstacle
$V = \|\mathbf{V}_{AO}\|$	Magnitude of the vehicle velocity relative to the obstacle
$\Delta\mathbf{V}_{AO}$	Vehicle acceleration relative to the obstacle
$\mathbf{V}_{AG} = \mathbf{V}_A - \mathbf{V}_G$	Vehicle velocity relative to the target G
$V_{AG} = \|\mathbf{V}_{AG}\|$	Magnitude of the vehicle's velocity relative to the target
$\Delta\mathbf{V}_{AG}$	Vehicle acceleration relative to the target
$\mathbf{L}_{AO}$	Vehicle position relative to the obstacle
$L = \|\mathbf{L}_{AO}\|$	Magnitude of the vehicle position relative to the obstacle
$\mathbf{L}_{AG}$	Vehicle position relative to target
$L_{AG} = \|\mathbf{L}_{AG}\|$	Magnitude of the vehicle position relative to the target
$P = \mathbf{V}_{AO}^{\mathrm{T}}\mathbf{L}_{AO}$	Temporal variable
$P_{AG} = \mathbf{V}_{AG}^{\mathrm{T}}\mathbf{L}_{AG}$	Temporal variable

Table 1. Nomenclatures used in this chapter (see Fig.1 and Fig. 2).

shown in Section 4.1 and 4.2, respectively. In Section 5, an application about the multiple task assignment (MTA) is used to verify the method proposed in this chapter. Finally, a brief conclusion is drawn in Section 6.

2. Relative velocity coordinates and the TPOA principles

2.1 Relative velocity coordinates

RVCs are constructed on the mass point of the vehicle as shown in Fig.1 and Fig.2. Fig.1 shows the relative coordinates when the obstacle-avoiding is considered while Fig.2 shows the scenario where the target-pursuing is considered. In these figures, the target G and obstacle O, which are 2D or 3D movers, are assumed as circles or spheres having certain radiuses which are denoted by R_{OS}. As mentioned above, the relative parameters are measurable or estimable for the current planning time, by using certain on-board sensors. For the convenience of description, the relative velocity and relative position between the vehicle A and the target G are denoted by V_{AG} and L_{AG}. Similarly, parameters about the obstacle O are denoted by V_{AO} and L_{AO}. Some other nomenclatures are listed in Table 1.

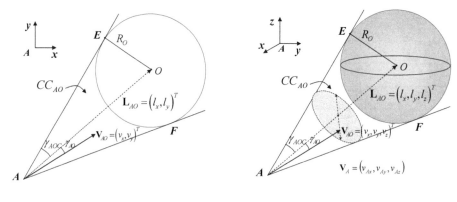

(a) Obstacle avoidance in two dimensions   (b) Obstacle avoidance in three dimensions

Fig. 1. Geometrical representation of the relative parameters in the relative coordinates when the obstacle-avoiding problem is considered. (a) and (b) show the definition of *relative obstacle angle* in 2D scenario and 3D scenario, respectively.

We define the *relative obstacle angle*, γ_{AO}, as the angle between $\mathbf{V}_{AO}$ and $\mathbf{L}_{AO}$. $\gamma_{AO} \in [0, \pi]$. *Collision cone* is defined as an area in which the vehicle will collide with an obstacle by current velocity $\mathbf{V}_{AO}$, and it is denoted by CC_{AO}. Generally, CC_{AO} is in the cone AEF, as shown in Fig.1. The half cone angle of CC_{AO}, γ_{AOC}, is defined as *collision region angle*, i.e.

$$\gamma_{AOC} = \arcsin\left(\frac{R_O}{L_{AO}}\right) \tag{1}$$

Similar to the definition in the obstacle-avoiding scenario in Fig.1, the angle between the relative velocity $\mathbf{V}_{AG}$ and relative distance $\mathbf{L}_{AG}$ is defined as *relative target angle*, denoted by $\gamma_{AG} \in [0, \pi]$, as shown in Fig.2. The *pursuit cone*, as well as the *pursuit region angle*, is defined in the same as obstacle avoidance scenario.

Some assumptions should be declared as the prerequisites of the planning principles. First, in order to guarantee the robot to catch the target successfully, the maximum velocity of the robot is assumed to be larger than that of the target and the obstacles. Due to this precondition, the vehicle A has the ability to catch the target, as well as avoid the collision with obstacles. Second, the target and the obstacles are supposed to keep their speed as constants during the planning period, τ. In fact, this hypothesis is usually accepted because τ is short enough for a real vehicle. Owe to the mechanism of numerical approximation, the path planning problem can be solved and optimized with a receding horizon fashion in

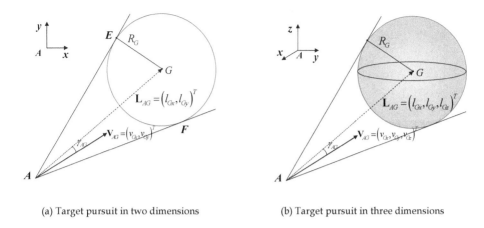

(a) Target pursuit in two dimensions   (b) Target pursuit in three dimensions

Fig. 2. Geometrical representation of the relative parameters in the relative coordinates when the target-pursuing problem is considered. (a) and (b) show the definition of *relative target angle* in 2D scenario and 3D scenario, respectively.

which a new waypoint of the total path is computed gradually over time. Here, we obtain the following equivalence.

$$\Delta \mathbf{V}_{AO} = \Delta \mathbf{V}_{AG} = \Delta \mathbf{V}_{A} \tag{2}$$

2.2 Obstacle-avoidance principle

For each obstacle O, under the assumption that its velocity is constant in τ, it will be avoided if the γ_{AO} is large enough to make the $\mathbf{V}_{AO}$ out of the CC_{AO} over the interval τ when the vehicle moves from time step k to $k+1$. This fact suggests that the obstacle-avoiding principle hold the following inequality. That is

$$\gamma_{AOC(k)} \leq \gamma_{AO(k+1)} \leq \pi \tag{3}$$

where $\gamma_{AOC(k)}$ is the *collision region angle* in time step k, which is shown in Fig.1. If there are multi obstacles, Eq. (3) changes to

$$\gamma_{AOCi(k)} \leq \gamma_{AOi(k+1)} \leq \pi \tag{4}$$

where the subscript i denotes the label of the obstacles. $i=1, 2\ldots N$. N stands for the number of obstacles.

2.3 Target-pursuit principle

The $\mathbf{V}_{AG}$ can be resolved into a pair of orthogonal components, as shown in Fig. 3. The vehicle is expected to tune its velocity to the optimum. Only when the velocity direction of the vehicle is identical to $\mathbf{L}_{AG}$, it will not lose its target. In the meanwhile, the vehicle should better minimize the magnitude of $\mathbf{V}_T$. Consequently, the policy for the optimal velocity is obvious in that $\mathbf{V}_T$ should be minimized while $\mathbf{V}_C$ maximized. We formulate these principles as two cost functions. They are

$$\min : V_T = \|\mathbf{V}_{AG}\| \sin \gamma_{AG} \tag{5}$$

and

$$\max : V_C = \|\mathbf{V}_{AG}\| \cos \gamma_{AG} \tag{6}$$

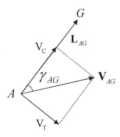

Fig. 3. The resolution of $\mathbf{V}_{AG}$. The velocity of the vehicle relative to the target is resolved into two orthogonal components, $\mathbf{V}_C$ and $\mathbf{V}_T$. One component, $\mathbf{V}_C$, is in the direction of $\mathbf{L}_{AG}$ and pointed to the target. The other, $\mathbf{V}_T$, is orthogonal to $\mathbf{L}_{AG}$.

3. Linear programming model

In this section, we first introduce some knowledge about the mathematical description of the standard linear programming model. Then, the mentioned principles for path planning are decomposed and linearized according to the linear prototype of the standard model.

3.1 Standard linear programming model
The standard Linear Programming is composed of a cost function and some constrains, all of which need to be affine and linear (Boyd & Vandenberghe, 2004). The following is the standard and inequality form.

$$\begin{aligned} \min_{\mathbf{x}} : &\mathbf{c}^T\mathbf{x} \\ \text{subject to:} &\mathbf{D}\mathbf{x} \le \mathbf{h} \end{aligned} \tag{7}$$

where $\mathbf{D} \in \mathbb{R}^{m \times n}$. The vector $\mathbf{x} \in \mathbb{R}^{n \times 1}$ denotes the variables. In the standard linear programming formulation, the vector $\mathbf{c} \in \mathbb{R}^{n \times 1}$, representing the coefficients, the vector $\mathbf{h} \in \mathbb{R}^{m \times 1}$, and the matrix $\mathbf{D}$ are all known when the model is going to be solved by some method.
During the decades, several methods have been invented to solve the optimization as shown in (7). From those methods, simplex algorithm and interior-point method are two most famous algorithms. More information about these algorithms can be found in (Boyd & Vandenberghe, 2004). In this chapter, an open library named Qsopt is used where the primal and dual simplex algorithms are imbedded. For the corresponding procedures, the user can specify the particular variant of the algorithm which is applied (David et al., 2011).

3.2 Linear programming model for path planning problem
In this subsection, the problem of obstacle-avoidance is formulated and we obtain some inequality constraints after we introduce two slack variables. Additionally, the formulation of

the target-pursuit problem is transferred to some cost functions. The final linear programming model consists of a weighted sum of the cost functions and a set of linear constraints. It is

$$\min : J = \sum_j \omega_j d_j + \omega_{v1} q_1 + \omega_{v2} q_2 \tag{8}$$

satisfying

$$\sum_j \omega_j + \omega_{v1} + \omega_{v2} = 1 \tag{9}$$

where ω_j, ω_{v1}, $\omega_{v2} \geq 0$. The letter j denotes the subscript of the components of $\mathbf{V}_{AG}$. If the path planning is operated in two-dimensional environment, j belongs to a two-element-set, i.e., $j \in \{x, y\}$. And if in three-dimensional environment, j belongs to a three-element-set, i.e., $j \in \{x, y, z\}$. The variables d_j, q_1, q_2 will be explained later.

3.2.1 Linearization of the avoidance constrain
From Fig.1, we get the *relative obstacle angle* function γ_{AO}.

$$\gamma_{AO} = \arccos\left(\frac{\mathbf{V}_{AO} \cdot \mathbf{L}_{AO}}{\|\mathbf{V}_{AO}\| \cdot \|\mathbf{L}_{AO}\|}\right) = \arccos\left(\frac{P}{VL}\right) \tag{10}$$

The definitions of V, L, and P refer to Table 1. Here, Eq. (10) is linearized by Taylor's theorem and we obtain the following equation.

$$
\begin{aligned}
\gamma_{AO(k+1)} &= \gamma_{AO(k)} + \Delta\gamma_{AO(k)} \\
&= \gamma_{AO(k)} + \tau\left(\frac{d\gamma_{AO}}{d\mathbf{V}_{AO}}\right)_{(k)} \frac{d\mathbf{V}_{AO}}{dt} + o\left(\|\Delta\mathbf{V}_{AO}\|\right)
\end{aligned} \tag{11}
$$

$$\Delta\gamma_{AO(k)} = -\frac{\tau}{\sqrt{V^2 L^2 - P^2}\Big|_{(k)}} \left(\mathbf{L}_{AO} - \frac{P}{V^2}\mathbf{V}_{AO}\right)_{(k)} \Delta\mathbf{V}_{AO} \tag{12}$$

Let $\gamma_{AO(k+1)}$ represent the *relative obstacle angle* in time step $k+1$ after vehicle's movement. The variables, $\Delta\mathbf{V}_{AO}$, in (11), are that we are trying to plan in step k for step $k+1$. If there are multiple obstacles, Eq. (12) changes to Eq. (13).

$$\Delta\gamma_{AOi(k)} = -\frac{\tau}{\sqrt{V^2 L^2 - P^2}\Big|_{i,(k)}} \left(\mathbf{L}_{AO} - \frac{P}{V^2}\mathbf{V}_{AO}\right)_{i,(k)} \Delta\mathbf{V}_{AO} \tag{13}$$

where

$$\gamma_{AOCi} = \arcsin\left(\frac{R_{Oi}}{L_i}\right) \tag{14}$$

3.2.2 Minimize the relative distance
In this subsection, the relative distance between the robot and the target-objective is minimized. By tuning the velocity of the robot, the relative distance between the robot

and target will became smaller through each step time. This distance is resolved into two or three elements by the axis number. That is to minimize the elements of the vector $\mathbf{L}_{AG}$, i.e.,

$$\text{min} : \left\| l_{Gj} - \left(v_{Gj}\tau + \Delta v_{Gj}\tau^2 \right) \right\| \tag{15}$$

where l_{Gj} is the element of $\mathbf{L}_{AG}$ and v_{Gj} is the element of $\mathbf{V}_{AG}$. See Fig.2. Δv_{Gj} denotes the acceleration of v_{Gj}. $j \in \{x, y\}$ is for two-dimensional environment and $j \in \{x, y, z\}$ is for three-dimensional environment. The inherited objective function is derived from Eq. (15).

$$\text{min} : d_j$$
$$\text{subject to} : -d_j \le l_{Gj} - \left(v_{Gj}\tau + \Delta v_{Gj}\tau^2 \right) \le d_j \tag{16}$$

where Δv_{Gj} is the variable that we hope to compute in the linear programming model, as stated in Eq. (2). In a manner of preserving convexity of the problem, $d_j \ge 0$ is the newly introduced slack variable associated with each element (Boyd & Vandenberghe, 2004).

3.2.3 Optimize the relative velocity
The relative velocity between the robot and the target-objective needs to optimize on the consideration of target-pursuit principle. We respectively discuss the optimization of the pair component, $\mathbf{V}_T$ and $\mathbf{V}_C$, as shown in Fig. 3.
(1) Minimize the magnitude of the component $\mathbf{V}_T$. We compute $\mathbf{V}_T$ with

$$V_T = \sqrt{V_{AG}^2 - \frac{P_{AG}^2}{L_{AG}^2}} \tag{17}$$

So the optimization of (5) is equal to

$$\text{min} : g\left(\bar{V}_{AG}\right) = V_T^2 = V_{AG}^2 - \frac{P_{AG}^2}{L_{AG}^2} \tag{18}$$

Using Taylor's theorem and introducing slack variable, q_1, we get the new formulation for optimization.

$$\text{min} : q_1$$
$$\text{subject to:} \quad 0 \le g\left(\mathbf{V}_{AG}\right) + \nabla g \Delta \mathbf{V}_{AG}^T \le q_1 \tag{19}$$
$$0 \le q_1 \le V_{AG\,\text{max}}^2$$

where $\Delta \mathbf{V}_{AG}$ is the variable vector that we hope to include in the linear programming model. q_1 is the slack variable. ∇g represents the grades of the function $g(.)$. It is computed with

$$\nabla g = 2\tau \left(\mathbf{V}_{AG} - \frac{P_{AG}}{L_{AG}^2} \mathbf{L}_{AG} \right) \tag{20}$$

$V_{AG\text{max}}$ is the maximum of the relative velocity between the robot and the target. We estimate it by $V_{AG\text{max}} = 2V_{A\text{max}}$. $V_{A\text{max}}$ denotes the upper boundary of V_A.

(2) Maximize the magnitude of the component $\mathbf{V}_C$.

Since we can maximize $\mathbf{V}_C$ by minimizing $-\mathbf{V}_C$, we refer to a minimize problem with affine cost function and constraint functions as a linear programming. Consequently, the problem described by (6) can be rewritten as

$$\min : -V_C = -\left\|\mathbf{V}_{AG}\right\|\cos\gamma_{AG} \tag{21}$$

Linearized using Taylor's theorem, Eq. (21) changes to a standard linear programming problem. That is

$$\min : q_2$$
$$\text{subject to: } -V_C - \nabla V_C \Delta \mathbf{V}_{AG}^T \leq q_2 \tag{22}$$
$$-V_{AG\max} \leq q_2 \leq V_{AG\max}$$

where $\Delta \mathbf{V}_{AG}$ is the variable that we are trying to include in our model. q_2 is also a slack variable. Here, the grades is computed with $\nabla \mathbf{V}_C = \tau \mathbf{L}_{AG}/L_{AG}$.

3.3 Other constraints

One of the advantages of the LP method is that various constraints can easily be added in the opening constraint set. Here we provide some demonstrations of constraints that are transformed from the dynamics, sensor data, and searching boundaries.

3.3.1 The constraints from the kinematics and dynamics

The kinematics and dynamics are extremely simplified in the path planning model where only the bound of the velocity and acceleration are added in the feasible set. These bound can be described mathematically as

$$\begin{cases} -V_{A\max} \leq v_{Aj} \leq V_{A\max} \\ -\Delta_{\max} \leq \Delta v_j \leq \Delta_{\max} \end{cases} \tag{23}$$

where v_{Aj} denotes the components of $\mathbf{V}_A$, as shown in Fig. 1. $\Delta_{\max}$ denotes the max magnitude of the changing of v_{Aj} in each period τ.

$$-V_{A\max} \leq v_{Aj} + \Delta v_j \tau \leq V_{A\max} \tag{24}$$

We get the net constraints (25) from the simultaneous equations (23) and (24).

$$\max\left\{-\Delta_{\max}, -\frac{1}{\tau}\left(V_{A\max} + v_{Aj}\right)\right\} \leq \Delta v_j \leq \min\left\{\Delta_{\max}, \frac{1}{\tau}\left(V_{A\max} - v_{Aj}\right)\right\} \tag{25}$$

3.3.2 The constraints from the limit of the sensor

All sensors can not detect the area that is beyond their capability. In real application, the obstacle that has been avoided by the vehicle will possibly disappear from the sensor's operating region. Similarly, any new obstacle from long distance is possible entering the detecting area. On this point, we evaluate the threat of each obstacle in the environment and propose a threat factor for each of them to estimate the performance. More specially, if the relative distance between the vehicle and the obstacle O_i satisfies $L_{AOi} \geq L_{mini}$, this obstacle is assigned $\lambda_i=0$. L_{mini} is the threshold. On the contrary, O_i is assigned $\lambda_i=1$ if $L_{AOi}<L_{mini}$.

In addition, the *relative obstacle angle* has impact on the vehicle. If the velocity vector of the vehicle satisfies $\gamma_{AOi} \geq \gamma_{AOCi} + \Delta\gamma_{AOimax}$, the vehicle can be looked as moving from the obstacle O_i, λ_i will be assigned zero.

3.3.3 The constraints from the searching region

The domain of the variables is generally computed in a square in 2D or a cube in 3D for the interceptive magnitude limit of the acceleration on each axis. In fact, the acceleration on each axis is coupled. So if the domain is treated as a circle in 2D (Richards & How, 2002) or a sphere in 3D, the algorithm will lower its conservation. See Fig. 4. We denote the domain as *dom*. That is

$$\Delta \mathbf{V}_A \in dom \tag{26}$$

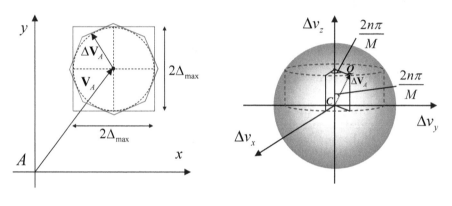

(a) Searching region in 2D;   (b) Searching region in 3D

Fig. 4. Searching region for acceleration. (a) The domain of the acceleration will be approximated by multiple lines in 2D. (b) The domain of the acceleration will be approximated by multiple planes in 3D.

We try to approximate the *dom* with multiple lines in 2D, or multiple planes in 3D. That is

$$\sin\frac{2m\pi}{M}\Delta v_x + \cos\frac{2m\pi}{M}\Delta v_y \leq \Delta_{max} \tag{27}$$

where $m=0,1,2,\ldots,$ M-1. Here, M represents the quantity of the line used for approximation. In three-dimensional environment,

$$\sin\frac{2m\pi}{M}\cos\frac{2n\pi}{M}\Delta v_x + \sin\frac{2m\pi}{M}\sin\frac{2n\pi}{M}\Delta v_y + \cos\frac{2m\pi}{M}\Delta v_z \leq \Delta_{max} \tag{28}$$

where $m, n=0,1,2,\ldots,M$-1. Here, M represents the quantity of the plane used for approximation. At this point, the linear programming model is formulized. The objective function is (8). All constrains are recapitulated as the following.

$$\begin{cases} \lambda_i \gamma_{AOCi} \le \lambda_i \left(\gamma_{AOi} + \Delta \gamma_{AOi} \right) \le \pi \\ -d_j \le l_j - \left(v_j \tau + \Delta v_j \tau^2 \right) \le d_j \\ 0 \le g\left(\mathbf{V}_{AG} \right) + \nabla g \cdot \Delta \mathbf{V}_{AG}^T \le q_1 \\ -V_C - \nabla \mathbf{V}_C \Delta \mathbf{V}_{AG}^T \le q_2 \\ \max\left\{ -\Delta_{max}, -\frac{1}{\tau} \left(V_{A\,max} + v_{Aj} \right) \right\} \le \Delta v_j \le \min\left\{ \Delta_{max}, \frac{1}{\tau} \left(V_{A\,max} - v_{Aj} \right) \right\} \\ \Delta \mathbf{V}_A \in dom \end{cases} \tag{29}$$

4. Path planning simulation for TPOA simulation

The approach proposed in this chapter is simulated with three obstacles and the results are given scenarios of in 2D and 3D, respectively. See Fig.5 and Fig.6. All the simulations run on the platform of WinXP/Pentium IV 2.53 GHz/2G RAM. A linear programming solver, named QSopt, is called from the library (David et al., 2011). We run the examples with three obstacles and give the results in two-dimensional environment and three-dimensional environment, respectively.

4.1 Path planning in 2D

According to the LP model, as shown with Eq. (29), all initial parameters are listed in Table 2. Assuming that the maximal velocity of the vehicle in 2D is 50cm/s, and the maximal acceleration of the vehicle is 350cm/s². The panning period is τ=20ms. The three parameters will be kept the same in the following simulation.

Fig. 5(a)~(d) show the key scenarios when avoiding the three obstacles. Fig. 5(a) shows the situation when the robot avoiding SO. It is obvious that the robot can rapidly adjust its velocity to the most favorable one and go on pursuing the target. The planner calls LP algorithm about 13 times in this stage. Fig. 5(b) shows the case while the robot has to avoid the moving obstacle MO1. At this time, the robot turns left to avoid MO1 because the over speeding is accessible for the robot. Fig. 5(c) shows the different decision that the robot selected to avoid MO2 from its back. Two conclusions can be drawn from this simulation. First, the robot is indeed qualified the avoiding and pursuing ability in uncertain environment. Second, the robot can adjust its velocity autonomously, including the magnitude and the direction, and adapt the optimal decision to avoid the obstacle and to catch the target. The whole pursuing process of this example lasts about 640ms and each period of calculation of our algorithm takes only about 0.469ms which is a low time complexity for real-time application.

	Initial Position (cm)	Initial Velocity (cm/s)	Radius (cm)
Robot (A)	(0,0)	(10,0)	N/A
Target (G)	(1000,1000)	(-12,0)	50
Static Obstacle (SO)	(300,300)	N/A	100
Moving Obstacle (MO1)	(800,600)	(-20,0)	50
Moving Obstacle (MO2)	(300,900)	(6,-3)	50

Table 2. The initial parameters for simulation in 2D.

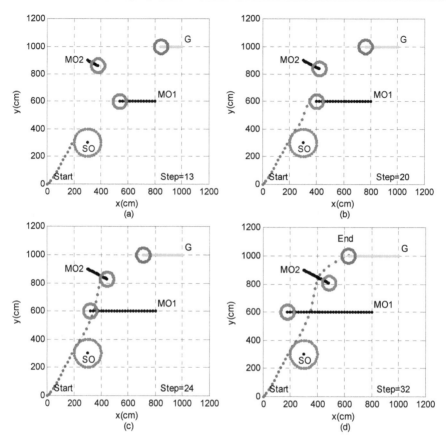

Fig. 5. Simulation in dynamic environment of 2D. The dot-line denotes the trajectories of the robot, the obstacles, and the target. And the circles with dot-line edge are the shape profile of any mover. SO represents static obstacle. MO1 and MO2 represent moving obstacles. G represents the moving target.

4.2 Path planning in 3D

For the specific 3D environment, assuming that there is one static obstacle SO1, and two moving obstacles, MO1 and MO2. All the initial parameters are listed in Table 3.

	Initial Position (cm)	Initial Velocity (cm/s)	Radius (cm)
Robot (A)	(0,1000,0)	(15,-15,0)	N/A
Target (G)	(1000,0,1000)	(-5,5,0)	50
Static Obstacle (SO)	(300,700,300)	N/A	100
Moving Obstacle (MO1)	(450,500,900)	(0,0,-10)	80
Moving Obstacle (MO2)	(450,250,850)	(0,10,0)	50

Table 3. The initial parameters for simulation in 3D.

Fig. 6 shows the results. At the beginning, SO is on the line between the robot and the target, and the robot is blocked. In the following time, the robot avoids MO1 and MO2 at step=19 and step=24, respectively. See Fig. 6(b) and (c). It is evident that the velocity-decision of the robot is optimized online. The time complexities of this simulation are 0.556ms in every period averagely.

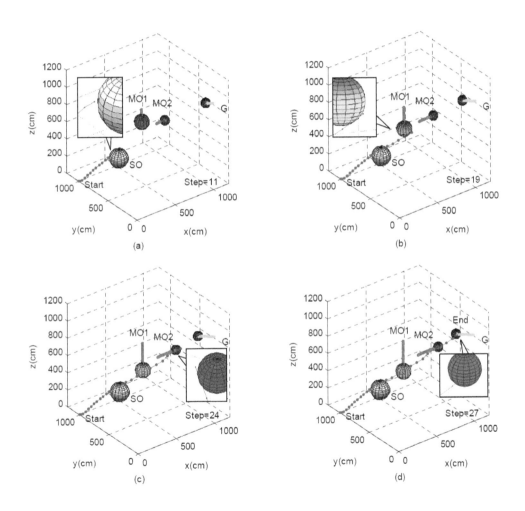

Fig. 6. Simulation in dynamic environment of 3D. The dot-line denotes the trajectories, and the sphere denotes the target and the obstacles.

5. Path planning for multiple tasks planning (MTP): an application

MTP problem is a variant of multiple TPOA. It can be stated that given N targets and N vehicles, let each vehicle can only pursue one target and each target is pursued exactly by one vehicle at any interval. The task is finished until all the targets have been caught. Then, the goal of MTP problem is to assign the tasks to vehicles within each interval online as well as complete the whole mission as fast as possible.

In this section, MTP problem is decomposed into two consecutive models, i.e., the task-assignment model and the path planning model. The path planning model has already been introduced above. With respect to the task-assignment model, a minimax assignment criterion is proposed to direct the task assignment. Under this assignment criterion, the optimal assignment will cost the least time to catch all the targets from the current measurable information.

Some simulations in Fig. 7 are given to show the efficiency of the method. Vehicles pursuing target are assigned according to the minimax assignment. According to the criterion, the current task assignment prefers to finish the whole mission fastest (Yang et al., 2009).

5.1 Assignment model under minimax assignment criterion

In the assignment problem, the payment in each vehicle for a target is assumed to be known. The problem is how to assign the target for each particular vehicle that the total pursuit mission can be completed as fast as possible.

Let x_{ij} be the n^2 0-1 decision variables, where $x_{ij}=1$ represents vehicle i for target j; otherwise, $x_{ij}=0$. Because the "payment" is understood as time, it is important to minimize the maximal time expended by vehicle for its pursuing process. This task assignment problem may be described as the minimax assignment model mathematically.

$$\min : z = \max_{i,j} \{c_{ij} x_{ij}\}$$

$$\text{subject to: } \sum_{i=1}^{n} x_{ij} = 1 \quad (j = 1,...,n)$$

$$\sum_{j=1}^{n} x_{ij} = 1 \quad (i = 1,...,n) \tag{30}$$

$$x_{ij} = 0 \text{ or } 1, \quad i, j \in \{1,...,n\}$$

where c_{ij} represents the payment in vehicle i for target j and will be given in Section 5.3. The elements c_{ij} give an n^2 binary cost matrix. This is a classic integer programming where the objective function is nonlinear and difficult to solve.

A solution method for the above minimax assignment problem named the operations on matrix is proposed. This solution method finds the solution directly from the cost matrix $C_n = (c_{ij})$ and the objective function (Yang et al., 2008).

5.2 Global cost function

We give the cost function of this MTP problem as the following. Let N_v, N_T, and N_o be the number of vehicles, targets and obstacles, respectively. The payment of the vehicle-i to pursue the target-j in the assignment problem is written as:

$$c_{ij} = \frac{d_{ij} + \xi_1 \sum_{k=1, k \neq i}^{N_v + N_o} d_{kij}}{\Delta v_{ij}} + \frac{\xi_2 |\Delta \Phi_{ij}|}{\omega_i} \qquad (31)$$

where d_{ij}, Δv_{ij} and $\Delta \Phi_{ij}$ are, respectively, the distance, velocity difference and heading difference between vehicle-i (V-i) and target-j (T-j). ω_i is the maximum turning rate of V-i; d_{kij} is the "additional distance" due to the obstacles and the vehicles other than V-i itself (with respect to vehicle-i, not only the obstacles, but also the other vehicles are its obstacles). ξ_1 and ξ_2 are positive constants. the c_{ij} in Eq. (33) indicates the time possibly needed by V-i to pursue T-j, and the "extended distance" consists of the linear distance, the angle distance, as well as the possible obstacle avoidance between V-i and T-j (Yang et al., 2009).

5.3 Simulation of MTP in 2D

The simulations demonstrated here include three robots, three moving target and three moving obstacles in two-dimensional environment (see Fig. 7). All the initial parameters are listed in Table 4.

	Initial Position (cm)	Initial Velocity (cm/s)	Radius (cm)
Robot (R1)	(0,200)	(17,0)	N/A
Robot (R2)	(0,0)	(17,0)	N/A
Robot (R3)	(200,0)	(17,0)	N/A
Target (G1)	(900,600)	(-11,9)	40
Target (G2)	(800,800)	(-10,10)	40
Target (G3)	(600,900)	(-9,11)	40
Moving Obstacle (MO1)	(850,450)	(-11,9)	60
Moving Obstacle (MO2)	(750,650)	(-10,10)	60
Moving Obstacle (MO3)	(550,750)	(-9,11)	60

Table 4. The initial parameters for MTP simulation in 2D.

In order to test the performance of the proposed linear programming model in the presence of different-motion vehicles, the following modifications are made during the pursuit process.
① at $k=0$, the maximal robot velocities are $v_{1\text{-max}}=30$ cm/s, $v_{2\text{-max}}=68$ cm/s, $v_{3\text{-max}}=45$ cm/s;
② at $k=16$, the maximal robot velocities are $v_{1\text{-max}}=30$ cm/s, $v_{2\text{-max}}=68$ cm/s, $v_{3\text{-max}}=22.5$ cm/s; ③ at $k=31$, the maximal robot velocities are $v_{1\text{-max}}=42.5$ cm/s, $v_{2\text{-max}}=68$ cm/s, $v_{3\text{-max}}=22.5$ cm/s.

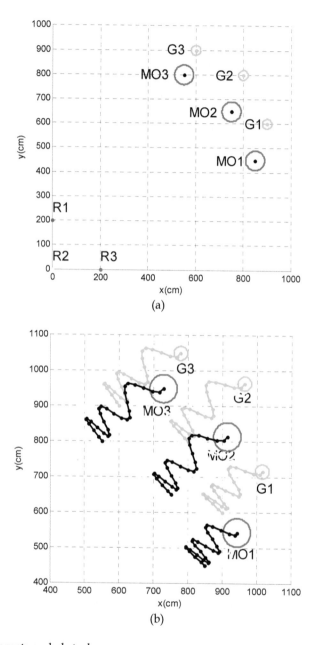

Fig. 7. Robots, targets and obstacles.
(a) Initial positions at *k*=0; (b) The target and obstacle trajectories till *k*=30. The targets and the obstacles are moving in a pso-sine curve.

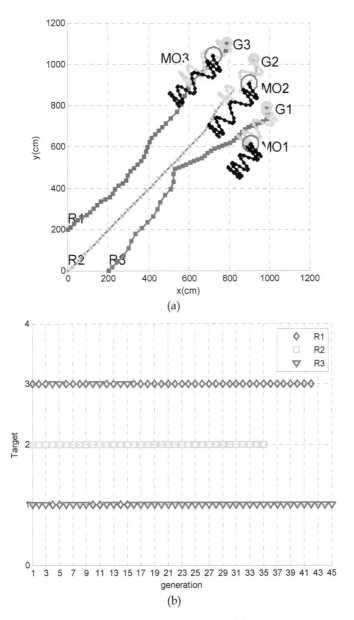

Fig. 8. Simulation of MTP in 2D. (a) Planning trajectories; (b) robot-target assignment pairs.

As Fig. 8 shows, the pair assignment alters between {R1-G3, R2-G2, R3-G1} and {R1-G1, R2-G2, R3-G3} at initial 15 steps according to the minimax assignment computation. At time step 16, assignment is kept for {R1-G3, R2-G2, R3-G1} due to that R3 is the slowest robot after its maximal velocity reducing 50%, and it can only catch G1. R2 is the fastest robot, so it pursues the fastest target G2. When $k=35$ and $k=42$, target G2 and G3 are caught by R2 and R1, respectively. Finally, G1 is caught at $k=45$ by R3, so the mission is completed successfully. The dot-lines are the planned trajectories of the robots and the colorful dots distinguish every step. From the figures we can see that the robots avoid the obstacles and catch all the moving targets successfully.

6. Conclusion

Path planning is a fundamental problem in robot application. In order to solve the path planning in dynamic environment, this chapter proposes a method based on LP/MILP to plan the acceleration of the robot in relative velocity coordinates. This method has the uniform formulation for 2D and 3D environment and it can give the information of the optimal velocity and acceleration in the view of the special cost function. Multiple constrains, such as the bounds of velocity, acceleration, and sensors, are included in the LP model and the MILP model. We also can add other constrains easily.

A particular application of this method is discussed for the problem of the multi-task planning where several robots are set to pursuit several targets. In the classical cooperation problem, the targets are assumed to be dynamic, similar to the TPOA problem. In the solution of MTP problem, a minimax assignment criterion and a global cost function are proposed to direct the task assignment.

Many simulations about the path planning in 2D/3D and in the multi-task planning requirements are taken to verify this novel method. The results show the low computing load of this method so it is potential to apply in real time manner.

7. Acknowledgment

This work was partially supported by Natural Science Foundation of China (NSFC) under grant #61035005 and #61075087, Hubei Provincial Education Department Foundation of China under grant #Q20111105, and Hubei Provincial Science and Technology Department Foundation of China under grant #2010CDA005.

8. References

Allaire, F. C. J.; Tarbouchi, M.; Labont´e G.; et al. (2009). FPGA implementation of genetic algorithm for UAV real-time path planning. *Journal of Intelligent and Robotic Systems*, Vol.54, No.1-3, pp.495-510, ISSN 09210296

Boyd S. & Vandenberghe L. (2004). Convex Optimization. Cambridge University Press, ISBN: 0521833787, New York

Chen, M.; Wu, Q. X. & Jiang, C. S. (2008). A modified ant optimization algorithm for path planning of UCAV. *Applied Soft Computing Journal*, Vol.8, No.4, pp.1712-1718, ISSN: 15684946

Chen, Y. & Han J. (2010). LP-Based Path Planning for Target Pursuit and Obstacle Avoidance in 3D Relative Coordinates. *Proceeding of American Control Conference*, pp.5394-5399, ISBN-13: 9781424474264, Baltimore, MD, USA, June 30- July 2, 2010

Chen, Yang; Zhao, Xingang; Zhang, Chan; et al. (2010). Relative coordination 3D trajectory generation based on the trimmed ACO. *Proceedings of International Conference on*

Electrical and Control Engineering, pp.1531-1536, ISBN-13: 9780769540313, Wuhan, China, June 26-28, 2010

Cocaud, C.; Jnifene, A. & Kim, B. (2008). Environment mapping using hybrid octree knowledge for UAV trajectory planning. *Canadian Journal of Remote Sensing*, Vol.34, No.4, pp.405-417, ISSN 07038992

David, A.; William, C.; Sanjeeb, D. & Monika, M. (2011). *QSopt Linear Programming Solver* [Online], June 1, 2011, available from:
<http://www2.isye.gatech.edu/~wcook/qsopt/index.html>

Fahimi, F.; Nataraj, C & Ashrafiuon, H. (2009). Real-time obstacle avoidance for multiple mobile robots. *Robotica*, Vol.27, No.2, pp.189-198, ISSN 02635747

Fiorini, P. & Shiller, Z. (1998). Motion Planning in Dynamic Environments Using Velocity Obstacles. *International Journal of Robotics Research*, Vol. 17, No.7, pp.760-772. ISSN 02783649

Hrabar, S. E. (2006). Vision-based 3D navigation for an autonomous helicopter, PhD thesis, University of Southern California, United States

Jung, L. F.; Knutzon, J. S.; Oliver, J. H, et al. (2006). Three-dimensional path planning of unmanned aerial vehicles using particle swarm optimization, *11th AIAA/ISSMO Multidisciplinary Analysis and Optimization Conference*, pp.992-1001, ISBN-10: 1563478234, Portsmouth, VA, USA, September 6-8, 2006

Kim, J. O. & Khosla, P. K. (1992). Real-time obstacle avoidance using harmonic potential functions. *IEEE Transactions on Robotics and Automation*, Vol.8, No.3, pp.338-349, ISSN 1042296X

Kitamura, Y.; Tanaka, T.; Kishino, F.; et al. (1995). 3-D path planning in a dynamic environment using an octree and an artificial potential field, *IEEE/RSJ International Conference on Intelligent Robots and Systems*, Vol.2, pp.474-481, Pittsburgh, PA, USA, August 5-9, 1995

Konolige, K. (2000). A gradient method for realtime robot control, *IEEE/RSJ International Conference on Intelligent Robots and Systems*, Vol.1, pp.639-646, Piscataway, NJ, USA, October 31-November 5, 2000

Krenzke, T. (2006). Ant Colony Optimization for agile motion planning. Master thesis. Massachusetts Institute of Technology, United States

Ladd, A. M. & Kavraki, L. E. (2004). Measure theoretic analysis of probabilistic path planning, *IEEE Transactions on Robotics and Automation*, Vol.20, No.2, pp.229-242, ISSN 1042296X

Liu, C.; Wei, Z. & Liu, C. (2007). A new algorithm for mobile robot obstacle avoidance based on hydrodynamics, *IEEE International Conference on Automation and Logistics*, pp.2310-2313, ISBN-10: 1424415314, Jinan, China, August 18-21, 2007

Nie, Y. Y.; Su, L. J. & Li, C. (2003). An Isometric Surface Method for Integer Linear Programming, *International Journal of Computer Mathematics*, Vol.80, No.7, pp.835-844, ISSN 00207160

Oh, T. S.; Shin, Y. S.; Yun, S. Y., et al. (2007). A feature information based VPH for local path planning with obstacle avoidance of the mobile robot, *4th International Conference on Mechatronics and Information Technology*. Vol.6794. Bellingham, WA, USA: SPIE

Richards, A. & How, J. P. (2002). Aircraft trajectory planning with collision avoidance using mixed integer linear programming. *Proceeding of American Control Conference*, pp. 936-1941, ISSN 07431619, Anchorage, AK, USA, May 8-10, 2002

Schouwenaars, T.; How, J. & Feron, E. (2004). Receding Horizon Path Planning with Implicit Safety Guarantees. *Proceeding of the American Control Conference*, pp. 5576-5581, ISSN 07431619, Boston, MA, USA, June 30 - July 2, 2004

Schumacher, C. J.; Chandler, P. R.; Pachter, M.; et al. (2004). Constrained optimization for UAV task assignment, *Proceedings of the AIAA guidance, navigation, and control conference*, AIAA 2004-5352, Providence, RI., 2004

Shima, T.; Rasmussen, S. J.; Sparks, A. G.; et al. (2006). Multiple task assignments for cooperating uninhabited aerial Vehicles using genetic algorithms, *Computers & Operations Research*, Vol.33, No.11, pp.3252-3269, ISSN 03050548

Wang, Y.; Lane, D. M. & Falconer, G. J. (2000). Two novel approaches for unmanned underwater vehicle path planning: constrained optimisation and semi-infinite constrained optimization. *Robotica*, Vol.18, No.2, pp.123-142, ISSN 02635747

Yang, L. Y.; Nie, M. H.; Wu, Z. W.; et al. (2008). Modeling and Solution for Assignment Problem, *International Journal of Mathematical Models and Methods in Applied Sciences*, Vol. 2, No. 2, pp.205-212, ISSN: 1998-0140

Yang, L. Y.; Wu, C. D.; Han, J. D.; et al. (2009). The task assignment solution for multi-target pursuit problem. *International Conference on Computational Intelligence and Software Engineering*, pp.1–6, ISBN-13 9781424445073, Wuhan, China, December 11-13, 2009

Zhang, Y. & Valavanis K P. (1997). A 3-D potential panel method for robot motion planning. *Robotica*, Vol.15, No.4, pp.421-434, ISSN 02635747

Zhao, L. & Murthy, V. R. (2007). Optimal flight path planner for an unmanned helicopter by evolutionary algorithms, *AIAA Guidance, Navigation and Control Conference*, Vol.4, pp.3716-3739, ISBN-10: 1563479044, Hilton Head, SC, USA, August 20-23, 2007

Zu, D.; Han, J. & Tan, D. (2006). Acceleration Space LP for the Path Planning of Dynamic Target Pursuit and Obstacle Avoidance. *Proceedings of the 6th World Congress on Intelligent Control and Automation*, Vol.2, pp.9084-9088, ISBN-10 1424403324, Dalian, China, June 21-23, 2006

8

Neural Networks Based Path Planning and Navigation of Mobile Robots

Valeri Kroumov[1] and Jianli Yu[2]
[1]Department of Electrical & Electronic Engineering,
Okayama University of Science, Okayama
[2]Department of Electronics and Information, Zhongyuan
University of Technology, Zhengzhou
[1]Japan
[2]China

1. Introduction

The path planning for mobile robots is a fundamental issue in the field of unmanned vehicles control. The purpose of the path planner is to compute a path from the start position of the vehicle to the goal to be reached. The primary concern of path planning is to compute *collision-free* paths. Another, equally important issue is to compute a *realizable* and, if possible, *optimal path*, bringing the vehicle to the final position.

Although humans have the superb capability to plan motions of their body and limbs effortlessly, the motion planning turns out to be a very complex problem. The best known algorithm has a complexity that is exponential to the number of degrees of freedom and polynomial in the geometric complexities of the robot and the obstacles in the environment (Chen & Hwang (1998)). Even for motion planning problems in the 2-dimensional space, existing complete algorithms that guarantee a solution often need large amount of memory and in some cases may take long computational time. On the other hand, fast heuristic algorithms may fail to find a solution even if it exists (Hwang & Ahuja (1992)).

In this paper we present a fast algorithm for solving the path planning problem for differential drive (holonomic)[1] robots. The algorithm can be applied to free-flying and snake type robots, too. Generally, we treat the two-dimensional known environment, where the obstacles are stationary polygons or ovals, but the algorithm can easily be extended for the three-dimensional case (Kroumov et al. (2010)). The proposed algorithm is, in general, based on the potential field methods. The algorithm solves the local minimum problem and generates optimal path in a relatively small number of calculations.

The paper is organized as follows. Previous work is presented in Section 2. In Section 3 we give a definition of the map representation and how it is used to describe various obstacles situated in the working environment. The path and the obstacle collisions are detected using artificial annealing algorithm. Also, the solution of the local minima problem is described there. In Section 4 we describe the theoretical background and the development of a motion

---

[1] In the mobile robotics, the term *holonomic* refers to the kinematic constraints of the robot chassis. The holonomic robot has zero nonholonomic kinematic constraints, while the nonholonomic one has one or more nonholonomic kinematic constraints.

planner for a differential drive vehicle. Section 5 presents the algorithm of the path planner. In Section 6 the calculation time is discussed and the effectiveness of the proposed algorithm is proven by presenting several simulation results. In the last section discussions, conclusions, and plans for further developments are presented.

2. Previous work

Comprehensive reviews on the work on the motion planning can be found in Latombe (1991). In this section we concentrate on motion planning for moving car-like robots in a known environment.

Motion planners can be classified in general as:

1) complete;

2) heuristic.

Complete motion planners can potentially require long computation times but they can either find a solution if there is one, or prove that there is none. Heuristic motion planners are fast but they often fail to find a solution even if it exists.

To date motion planners can be classified in four categories (Latombe (1991)):

1) skeleton;

2) cell decomposition;

3) subgoal graph;

4) potential field.

In the skeleton approach the free space is represented by a network of one-dimensional paths called a *skeleton* and the solution is found by first moving the robot onto a point on the skeleton from the start configuration and from the goal, and by connecting the two points via paths on the skeleton. The approach is intuitive for two-dimensional problems, but becomes harder to implement for higher degrees of freedom problems.

Algorithms based on the visibility graph (VGRAPH) (Lozano-Pèrez & Wesley (1979)), the Voronoi diagram (O'Dúnlaing & Yap (1982)), and the *silhouette* (Canny (1988)) (projection of obstacle boundaries) are examples of the skeleton approach. In the VGRAPH algorithm the path planning is accomplished by finding a path through a graph connecting vertices of the forbidden regions (obstacles) and the generated path is near optimal. The drawback of the VGRAPH algorithm is that the description of all the possible paths is quite complicated. Actually, the number of the edges of the VGRAPH is proportional to the squared total number of the obstacles vertices and when this number increases, the calculation time becomes longer. Another drawback is that the algorithm deals only with polygonal objects. Yamamoto et al. (1998) have proposed a near-time-optimal trajectory planning for car-like robots, where the connectivity graph is generated in a fashion very similar to that of Lozano-Pèrez & Wesley (1979). They have proposed optimization of the speed of the robot for the generated trajectory, but the optimization of the trajectory itself is not enough treated.

In the cell decomposition approach (Paden et al. (1989)) the free space is represented as a union of cells, and a sequence of cells comprises a solution path. For efficiency, hierarchical trees, e.g. octree, are often used. The path planning algorithm proposed by Zelinsky (1992) is quite reliable and combines some advantages of the above algorithms. It makes use of quadtree data structure to model the environment and uses the distance transform methodology to generate paths. The obstacles are polygonal-shaped which yields a quadtree with minimum sized leaf quadrants along the edges of the polygon, but the quadtree is large and the number of leaves in the tree is proportional to the polygon's perimeter and this makes it memory consuming.

In the subgoal-graph approach, subgoals represent *key* configurations expected to be useful for finding collision-free paths. A graph of subgoals is generated and maintained by a global planner, and a simple local planner is used to determine the reachability among subgoals. This two level planning approach is first reported by Faverjon & Tournassoud (1987) and has turned out to be one of the most effective path planning methods.

The potential field methods and their applications to path planning for autonomous mobile robots have been extensively studied in the past two decades (Khatib (1986), Warren (1989), Rimon & Doditschek (1992), Hwang & Ahuja (1992), Lee & Kardaras (1997a), Lee & Kardaras (1997b), Chen & Hwang (1998), Ge & Cui (2000), Yu et al. (2002), Kroumov et al. (2004)). The basic concept of the potential field methods is to fill the workspace with an artificial potential field in which the robot is attracted to the goal position being at the same time repulsed away from the obstacles (Latombe (1991)). It is well known that the strength of potential field methods is that, with some limited engineering, it is possible to construct quite efficient and relatively reliable motion planners (Latombe (1991)). But the potential field methods are usually incomplete and may fail to find a free path, even if one exists, because they can get trapped in a local minimum (Khosla & Volpe (1988); Rimon & Doditschek (1992); Sun et al. (1997)). Another problem with the existing potential field methods is that they are not so suitable to generate optimal path: adding optimization elements in the algorithm, usually, makes it quite costly from computational point of view (Zelinsky (1992)). Ideally, the potential field should have the following properties (Khosla & Volpe (1988)):

1. The magnitude of the potential field should be unbounded near obstacle boundaries and should decrease with range.

2. The potential should have a spherical symmetry far away from the obstacle.

3. The equipotential surface near an obstacle should have a shape similar to that of the obstacle surface.

4. The potential, its gradient and their effects on the path must be spatially continuous.

The proposed in this paper algorithm is partially inspired by the results presented by Sun et al. (1997) and by Lee & Kardaras (1997a), but our planner has several advantages:

- the obstacle descriptions are implemented directly in the simulated annealing neural network—there is no need to use approximations by nonlinear equations (Lee & Kardaras (1997b));

- there is no need to perform a learning of the workspace *off-line* using backpropagation (Tsankova (2010));

- there is no need to perform additional calculations and processing when an obstacle is added or removed;

- a simple solution of the local minimum problem is developed and the generated paths are conditionally optimized in the sense that they are piecewise linear with directions changing at the corners of the obstacles.

The algorithm allows parallel calculation (Sun et al. (1997)) which improves the computational time. The experimental results show that the calculation time of the presented algorithm depends linearly on the total number of obstacles' vertices—a feature

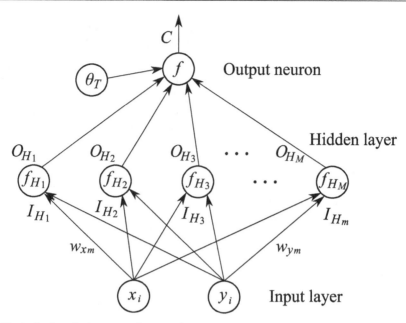

Fig. 1. Obstacle description neural network

which places it among the fastest ones. The only assumptions made in this work are that there are a finite number of stationary oval or polygonal obstacles with a finite number of vertices, and that the robot has a cylindrical shape. The obstacles can be any combination of polygons and ovals, as well. In order to reduce the problem of path planning to that of navigating a point, the obstacles are enlarged by the robot's polygon dimensions to yield a new set of polygonal obstacles. This "enlargement" of the obstacles is a well known method introduced formally by Lozano-Pèrez & Wesley (1979).

3. Environment description

3.1 Obstacles
Every obstacle is described by a neural network as shown in Fig. 1. The inputs of the networks are the coordinates of the points of the path. The output neuron is described by the following expression, which is called *a repulsive penalty function (RPF)* and has a role of repulsive potential:

$$C = f(I_0) = f\left(\sum_{m=1}^{M} O_{H_m} - \theta_T\right), \tag{1}$$

where I_0 takes a role of the induced local field of the neuron function $f(\cdot)$, θ_T is a bias, equal to the number of the vertices of the obstacle decreased by 0.5. The number of the neurons in the hidden layer is equal to the number of the vertices of the obstacle. O_{H_m} in eq. (1) is the output of the m-th neuron of the middle layer:

$$O_{H_m} = f(I_{H_m}), \quad m = 1, \ldots, M, \tag{2}$$

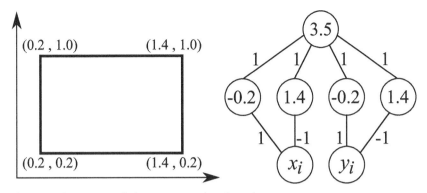

Fig. 2. The annealing network for a rectangular obstacle

where I_{H_m} is the weighted input of the m-th neuron of the middle layer and has a role of induced local field of the neuron function. The neuron activation function $f(\cdot)$ has the form:

$$f(x) = \frac{1}{1 + e^{-x/T}},\tag{3}$$

where T is the pseudotemperature and the induced local field (x) of the neuron is equal to I_0 for eq. (1) or equal to I_{H_m} in the case of eq. (2). The pseudotemperature decreasing is given by:

$$T(t) = \frac{\beta_0}{\log(1 + t)}.\tag{4}$$

Finally, I_{H_m} is given by the activating function

$$I_{H_m} = w_{xm}x_i + w_{ym}y_i + \theta_{H_m},\tag{5}$$

where x_i and y_i are the coordinates of i-th point of the path, w_{xm} and w_{ym} are weights, and θ_{H_m} is a bias, which is equal to the free element in the equation expressing the shape of the obstacle.

The following examples show descriptions of simple objects.

Example: 1. *Description of polygonal obstacle (see Fig. 2).*

The area inside the obstacle can be described with the following equations:

$$x - 0.2 > 0; \qquad -x + 1.4 > 0$$
$$y - 0.2 > 0; \qquad -y + 1.0 > 0.$$

From these equations and eq. (5):

$$w_{x1} = 1 \quad w_{y1} = 0 \quad \theta_{H_1} = -0.2$$
$$w_{x2} = -1 \quad w_{y2} = 0 \quad \theta_{H_2} = 1.4$$
$$w_{x3} = 0 \quad w_{y3} = 1 \quad \theta_{H_3} = -0.2$$
$$w_{x4} = 0 \quad w_{y4} = -1 \quad \theta_{H_4} = 1$$

i.e. in the middle layer of the network, the activating functions become

$$I_{H_1} = w_{x1}x_i + w_{y1}y_i + \theta_{H_1} = x_i - 0.2$$
$$I_{H_2} = w_{x2}x_i + w_{y2}y_i + \theta_{H_2} = -x_i + 1.4$$
$$I_{H_3} = w_{x3}x_i + w_{y3}y_i + \theta_{H_3} = y_i - 0.2$$
$$I_{H_4} = w_{x4}x_i + w_{y4}y_i + \theta_{H_4} = -y_i + 1.0.$$

Hence, the free elements in the equations describing the obstacle are represented by the biases θ_{H_i} and the weights in the equations for I_{H_i} are the coefficients in the respective equations.

Example: 2. *Circular shape obstacle.*

When an obstacle has a circular shape, I_{H_m} is expressed as:

$$I_H = R^2 - (x_i - P)^2 - (y_i - Q)^2, \tag{6}$$

where R is the radius and (P, Q) are the coordinates of the centre.

Example: 3. *Description of elliptical obstacles.*

When the obstacle has elliptic (circular) shape, I_{H_m} is expressed as:

$$I_H = a^2b^2 - (X - x_i)^2b^2 + (Y - y_i)^2a^2, \tag{7}$$

which comes directly from the standard equation of the ellipse

$$\frac{(X - x_i)^2}{a^2} + \frac{(Y - y_i)^2}{b^2} = 1,$$

with xy-coordinates of the foci $(-\sqrt{a^2 - b^2} + X, Y)$ and $(\sqrt{a^2 - b^2} + X, Y)$ respectively, and the description network has two neurons in the middle layer.

Note 1. *The oval shaped obstacles are represented by neural network having two neurons (see the above Example 2 and Example 3) in the middle layer.*

It is obvious from the above that any shape which can be expressed mathematically can be represented by the description neural network. This, of course, includes configurations which are a combination of elementary shapes.

The description of the obstacles by the shown here network has the advantage that it can be used for parallel computation of the path, which can increase the speed of path generation. As it will be shown later, this description of the obstacles has the superiority that the calculation time depends only on the total number of the obstacles' vertices.

3.2 The local minima problem

The local minima remain an important cause of inefficiency for potential field methods. Hence, dealing with local minima is the major issue that one has to face in designing a planner based on this approach. This issue can be addressed at two levels (Latombe (1991)): (1) definition of the potential function, by attempting to specify a function with no or few local minima, and (2) in the design of the search algorithm, by including appropriate techniques for escaping from local minima. However, it is not easy to construct an "ideal" potential function with no local minima in a general configuration. The second level is more realistic and is addressed by many researchers (see e.g. Latombe (1991); Lozano-Pèrez et al. (1994) and the references there).

In the proposed in this chapter algorithm, the local minima problem is addressed in a simple and efficient fashion:

1. After setting the coordinates of the start and the goal points respectively (Fig. 3(a)), the polygonal obstacles are scanned in order to detect concavity (Fig. 3(b)). In the process of scanning, every two other vertices of a given obstacle are connected by a straight line and if the line lies outside the obstacle, then a concavity exists.

2. If the goal (or the start) point lies inside a concavity, then a new, temporary, goal (start) lying outside the concavity is set (point g' in Fig. 3(b)), and the initial path between the original goal (start) and the new one is set as a straight line. The temporary goal (start) is set at the nearest to the "original" start (goal) vertex.

3. Every detected concavity is temporarily filled, i.e. the obstacle shape is changed from a concave to a nonconcave one (see Fig. 3(c)). After finishing the current path-planning task, the original shapes are retained so that the next task could be planned correctly, and the temporary goal (start) is connected to the original one by a straight line (Fig. 3(d)).

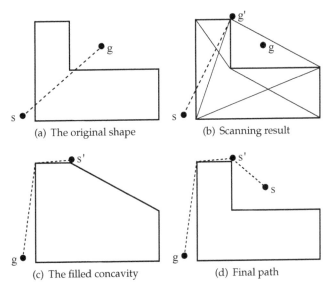

Fig. 3. Concave obstacle (the local minima solution)

The above allows isolating the local minima caused by concavities and increases the reliability of the path planning.

4. Optimal path planner

The state of the path is described by the following energy function.

$$E = w_l E_l + w_c E_c, \tag{8}$$

where w_l and w_c are weights ($w_l + w_c = 1$), E_l depicts the squared length of the path:

$$E_l = \sum_{i=1}^{N-1} L_i^2 = \sum_{i=1}^{N-1} [(x_{i+1} - x_i)^2 + (y_{i+1} - y_i)^2], \tag{9}$$

and E_c is given by the expression.

$$E_c = \sum_{i=1}^{N} \sum_{k=1}^{K} C_i^k, \tag{10}$$

where N is the number of the points between the start and the goal, K is the number of the obstacles, and C is obtained through eq. (1).

Minimizing eq. (8) will lead to obtaining an optimal in length path that does not collide with any of the obstacles. In order to minimize (8) the classical function analysis methods are applied. First, we find the time derivative of E:

$$\frac{dE}{dt} = \sum_{i=1}^{N} \left[w_l \left(\frac{\partial L_i^2}{\partial x_i} + \frac{\partial L_{i-1}^2}{\partial x_i} \right) + w_c \sum_{k=1}^{K} \frac{\partial C_i^k}{\partial x_i} \right] \dot{x}_i$$

$$+ \sum_{i=1}^{N} \left[w_l \left(\frac{\partial L_i^2}{\partial y_i} + \frac{\partial L_{i-1}^2}{\partial y_i} \right) + w_c \sum_{k=1}^{K} \frac{\partial C_i^k}{\partial y_i} \right] \dot{y}_i \tag{11}$$

Setting

$$\dot{x}_i = -\eta \left[w_l \left(\frac{\partial L_i^2}{\partial x_i} + \frac{\partial L_{i-1}^2}{\partial x_i} \right) + w_c \sum_{k=1}^{K} \frac{\partial C_i^k}{\partial x_i} \right],$$

$$\dot{y}_i = -\eta \left[w_l \left(\frac{\partial L_i^2}{\partial y_i} + \frac{\partial L_{i-1}^2}{\partial y_i} \right) + w_c \sum_{k=1}^{K} \frac{\partial C_i^k}{\partial y_i} \right] \tag{12}$$

we can rewrite the above equation as

$$\frac{dE}{dt} = -\frac{1}{\eta} \sum_{i=1}^{N} (\dot{x}_i^2 + \dot{y}_i^2) < 0 \tag{13}$$

where η is an adaptation gain. It is obvious that, when $\dot{x}_i \rightarrow 0$ and $\dot{y}_i \rightarrow 0$, E converges to its minimum. In other words, when all points of the path almost stop moving, there is no collision and the path is the optimal one, i.e. eq. (13) can be used as a condition for termination of the calculation iterations.

Now, from equations (12),

$$\frac{\partial L_i^2}{\partial x_i} + \frac{\partial L_{i-1}^2}{\partial x_i} = -2x_{i+1} + 4x_i - 2x_{i-1},$$

$$\frac{\partial L_i^2}{\partial y_i} + \frac{\partial L_{i-1}^2}{\partial y_i} = -2y_{i+1} + 4y_i - 2y_{i-1} \tag{14}$$

and

$$\frac{\partial C_i^k}{\partial x_i} = \frac{\partial C_i^k}{\partial (I_0)_i^k} \frac{\partial (I_0)_i^k}{\partial x_i} = \frac{\partial C_i^k}{\partial (I_0)_i^k} \left(\sum_{m=1}^{M} \frac{\partial (O_{H_m})_i^k}{\partial (I_{H_m})_i^k} \frac{\partial (I_{H_m})_i^k}{\partial x_i} \right)$$

$$= f'((I_0)_i^k) \left(\sum_{m=1}^{M} f'_{H_m}((I_{H_m})_i^k) w_{xm}^k \right);$$

$$\frac{\partial C_i^k}{\partial y_i} = \frac{\partial C_i^k}{\partial (I_0)_i^k} \frac{\partial (I_0)_i^k}{\partial y_i} = \frac{\partial C_i^k}{\partial (I_0)_i^k} \left(\sum_{m=1}^{M} \frac{\partial (O_{H_m})_i^k}{\partial (I_{H_m})_i^k} \frac{\partial (I_{H_m})_i^k}{\partial x_i} \right)$$

$$= f'((I_0)_i^k) \left(\sum_{m=1}^{M} f'_{H_m}((I_{H_m})_i^k) w_{ym}^k \right). \tag{15}$$

This leads to the final form of the function:

$$\dot{x}_i = -2\eta w_l(2x_i - x_{i-1} - x_{i+1}) - \eta w_c \sum_{k=1}^{K} f'((I_0)_i^k) \left(\sum_{m=1}^{M} f'_{H_m}((I_{H_m})_i^k) w_{xm}^k \right);$$

$$\dot{y}_i = -2\eta w_l(2y_i - y_{i-1} - y_{i+1}) - \eta w_c \sum_{k=1}^{K} f'((I_0)_i^k) \left(\sum_{m=1}^{M} f'_{H_m}((I_{H_m})_i^k) w_{ym}^k \right), \tag{16}$$

where f' is given by the following expressions:

$$f'(\cdot) = \frac{1}{T} f(\cdot)[1 - f(\cdot)]$$

$$f'_{H_m}(\cdot) = \frac{1}{T} f_{H_m}(\cdot)[1 - f_{H_m}(\cdot)]. \tag{17}$$

In eq. (16) the first member in the right side is for the path length optimization and the second one is for the obstacle avoidance.

One of the important advantages of the above path-planning is that it allows parallelism in the calculations of the neural network outputs (see Section 6), which leads to decreasing the computational time. The generated semi-optimal path can be optimized further by applying evolutionary methods (e.g. genetic algorithm). Such approach leads to an optimal solution but the computational cost increases dramatically (Yu et al. (2003)).

5. The path-planning algorithm

In this section an algorithm based on the background given in Sections 3 and 4 is proposed. The calculations for the path are conceptually composed by the following steps:

Step 1: Initial step

1. Let the start position of the robot is (x_1, y_1), and the goal position is denoted as (x_N, y_N).
2. Check for concavities (see Section 3.2.) and, if necessary, reassign the goal (start) position.
3. At $t = 0$ the coordinates of the points of the initial path (straight line) $(x_i, y_i; \ i = 2, 3, \ldots, N-1)$ are assigned as

$$x_i = x_0 + i(x_N - x_1)/(N-1),$$

$$y_i = (y_N - y_1)(x_i - x_1)/(x_N - x_1) + y_0, \tag{18}$$

i. e. the distance between the x and y coordinates of every two neighboring points of the path is equal.

Note 2. *It is assumed that the obstacles dimensions are enlarged by the robot's polygon dimensions (Lozano-Pèrez & Wesley (1979)).*

Step2: 1. For the points (x_i, y_i) of the path which lie inside some obstacle, the iterations are performed according to the following equations:

$$\dot{x}_i = -2\eta_1 w_l(2x_i - x_{i-1} - x_{i+1}) - \eta_1 w_c \sum_{k=1}^{K} f'((I_0)_i^k) \left(\sum_{m=1}^{M} f'_{H_m}((I_{H_m})_i^k) w_{xm}^k \right);$$

$$\dot{y}_i = -2\eta_1 w_l(2y_i - y_{i-1} - y_{i+1}) - \eta_1 w_c \sum_{k=1}^{K} f'((I_0)_i^k) \left(\sum_{m=1}^{M} f'_{H_m}((I_{H_m})_i^k) w_{ym}^k \right) \quad (19)$$

$$i = 2, 3, \ldots, N - 1.$$

2. For the points (x_i, y_i) situated outside the obstacles, instead of eq. (19) use the following equations:

$$\dot{x}_i = -\eta_2 w_l(2x_i - x_{i-1} - x_{i+1});$$
$$\dot{y}_i = -\eta_2 w_l(2y_i - y_{i-1} - y_{i+1}), \quad (20)$$

i.e. for the points of the path lying outside obstacles, we continue the calculation with the goal to minimize only the length of the path.

Step 3: Perform p times the calculations of step 2, i.e. find $x_i(t+p), y_i(t+p)$ $(i = 2, 3, \ldots, N-1)$, where p is any suitable number, say $p = 100$.

Step 4: Test for convergence

Calculate the difference d of the path lengths at time t and time $(t+p)$, i.e.

$$d = \sum_{i=2}^{N-1} \{[x_i(t+p) - x_i(t)]^2 + [y_i(t+p) - y_i(t)]^2\}^{1/2}. \quad (21)$$

- If $d < \varepsilon$ then the algorithm terminates with the conclusion that the goal is reached via an "optimal" path. Here ε is a small constant, say $\varepsilon = 0.1$.
- If $d \geq \varepsilon$, then GO TO step 2.

Here eq. (19) is almost the same as eq. (16) with the difference that instead of only one RPF, different functions, as explained below, are used for every layer of the neural network. Every obstacle is described using a neural network as shown in Fig. 1. The output neuron is described by eq. (1), the neuron function $f(\cdot)$ is the same as in eq. (3) and the pseudotemperature is as in eq. (4).

The hidden layer inputs are as in eq. (1) but the outputs O_{H_m} now become

$$O_{H_m} = f_{H_m}(I_{H_m}), \quad m = 1, \ldots, M \quad (22)$$

with I_{H_m} becoming the induced local field of the neuron function f_{H_m}:

$$f_{H_m}(I_{H_m}) = \frac{1}{1 + e^{-I_{H_m}/T_{H_m}(t)}} \quad (23)$$

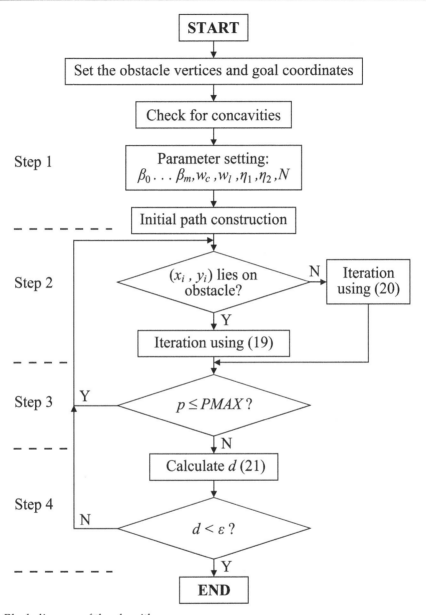

Fig. 4. Block diagram of the algorithm

and

$$T_{H_m}(t) = \frac{\beta_m}{\log(1+t)} \tag{24}$$

Finally, I_{H_m} is given by the activating function (5). Equations (3) and (23) include different values of pseudotemperatures, which is one of the important conditions for convergence of the algorithm and for generation of almost optimal path. The block diagram of the algorithm is shown in Fig. 4.

6. Experimental results

To show the effectiveness of the proposed in this paper algorithm, several simulation examples are given in this section.

6.1 Calculation speed evaluation

The simulations were run on ASUS A7V motherboards rated with 328 base (352 peak) SPEC CFP2000 and 409 base (458 peak) SPEC CINT2000 on the CPU2000 benchmark. The environment shown in Fig. 5 has been chosen for the benchmark tests. In the course of the speed measurements the number of the vertices was increased from 20 to 120 with step of 20 vertices. The final configuration of the obstacles inside the benchmark environment with three generated paths is shown in Fig. 5. To compare the run-times, 12 path generations have been performed randomly. After deleting the lowest and the highest path planning times, the average of the remaining 10 values has been adopted.

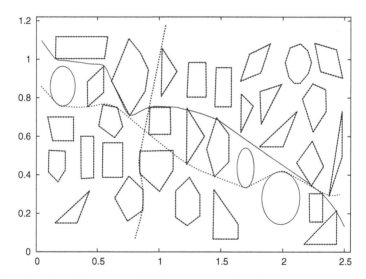

Fig. 5. The benchmark environment with three generated paths

Figure 6 shows the speed change depending on the total number of vertices of the obstacles. It is clear that the speed changes linearly with increasing the number of the vertices (cf. Lozano-Pèrez & Wesley (1979)).

As it was explained at the end of Section 4 the algorithm allows parallelizing the path computation. To calculate the speedups we ran parallel simulations for the same benchmark environment on one, two, and three processors. Figure 7 shows how the speed increases with increasing the number of the processors. For the case of two processors the simulation was run on two ASUS A7V boards, and for the case of three processors a Gigabyte 440BX board was added as a third processor. The communication during the parallel calculations was established by an Ethernet bus based network between the computers.

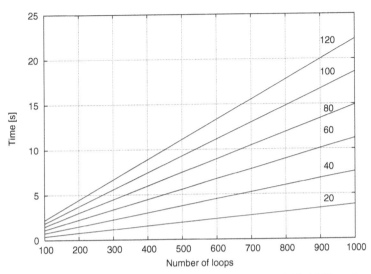

Fig. 6. Computation time depending on the number of the loops (100–1000) and number of the vertices (20–120)

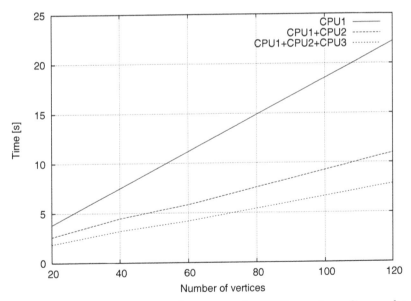

Fig. 7. Run-time t for different number of processors for 1000 loops depending on the number of vertices

It is possible to further increase the calculation speed by introducing adaptive adjustment of the parameters η_1 and η_2. For example, the following adaptive adjusting law increases the speed about three times (see Fig. 8):

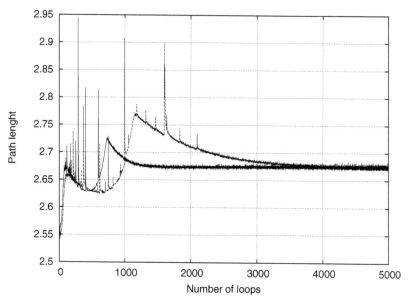

Fig. 8. Convergence speed of the algorithm depending on the way of setting η_1 and η_2: the slowly converging line is for the case when constant values are used, and the fast converging one depicts the convergence when adaptive adjustment is applied

$$\dot{\eta}_1(n+1) = -\eta_1(n) + \delta \, \mathrm{sgn}[(d(n) - d(n-1))(d(n-1) - d(n-2))],$$
$$\dot{\eta}_2(n+1) = -\eta_2(n) + \delta \, \mathrm{sgn}[(d(n) - d(n-1))(d(n-1) - d(n-2))] \qquad (25)$$

where δ is a small constant.

In the course of path calculations, a limited number of obstacles has impact on path generation—mostly the obstacles the initial, straight line path intersects with, contribute to path forming. Then, while performing *Step 1* of the algorithm, if all obstacles which do not collide with the initial path are ignored and excluded from the calculations, it is possible to drastically decrease the calculation time. A good example is the path in vertical direction in Fig. 5. When all obstacles are considered, the total number of vertices is 135 and RPF for all obstacles would be calculated. However, if only the four obstacles which collide with the initial straight line are used, the number of vertices decreases to only 17 and the expected calculation time will be approximately 8 times shorter. After finishing the path generation for the decreased number of obstacles, the original environment should be retained and a final correction of the path shape should be performed. Naturally, this final correction would require only few additional calculation loops.

6.2 Simulation results

Figure 9 shows additional results of path planning simulation in an environment with concaved obstacles. It can be seen that the local minimums do not cause problems in the path generation. In these simulations the number of the points between the start position and the goal was set to 200. In both examples the values of the coefficients in equations of the algorithm were not changed. In fact, the authors have performed many simulations using different environments, in which the obstacles were situated in many different ways. In all

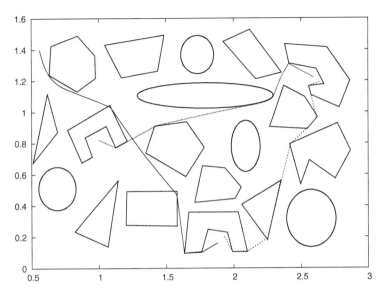

Fig. 9. Simulation example (environment including concave obstacles)

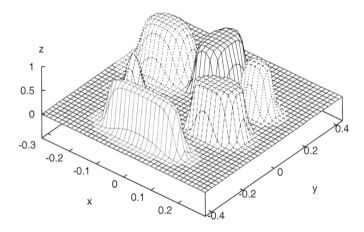

Fig. 10. Shape of the RPF in the course of simulation

the simulations the paths were successfully generated and the time for the calculations agreed with the calculation speed explained above. Figure 10 shows the potential fields of obstacles in the course of a simulation.

We have developed software with a graphic user interface (GUI) for the simulation and control of differential drive robots in a 2D environment (See Fig. 11). The environment allows the user to easily build configuration spaces and perform simulations. This environment is available at *http://shiwasu.ee.ous.ac.jp/planner/planner.zip*. The algorithm was successfully applied to the

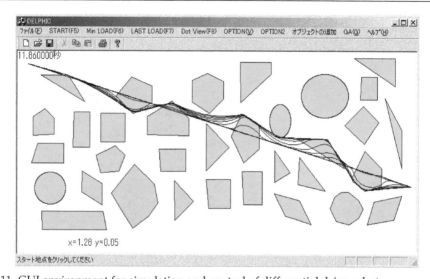

Fig. 11. GUI environment for simulation and control of differential drive robots

control and navigation of "Khepera" (Mondada et al. (1993)) and "Pioneer P3-DX" robots.

7. Discussion and conclusions

In this paper we have proposed an algorithm which guarantees planning of near optimal in length path for wheeled robots moving in an *a priori* known environment. We are considering wheel placements which impose *no* kinematic constraints on the robot chassis: castor wheel, Swedish wheel, and spherical wheel.

The proposed algorithm is a significant improvement of the potential field algorithms because it finds an almost optimal path without being trapped in local minima and the calculation speed for the proposed algorithm is reasonably fast. Because the only assumptions made are that the obstacles are stationary polygons and ovals, the algorithm can solve the navigation problem in very complex environments. For description of the obstacles the algorithm uses only the Cartesian coordinates of their vertices and does not need memory consuming obstacle transformations (cf. Zelinsky (1992); Zelinsky & Yuta (1993)). The consumed memory space is equal to twice the number of the coordinates of the path points and some additional memory for keeping the parameters of the description networks.

Because the generated paths are piecewise linear with changing directions at the corners of the obstacles, the inverse kinematics problems for the case of differential drive robots are simply solved: to drive the robot to some goal pose (x, y, θ), the robot can be spun in place until it is aimed at (x, y), then driven forward until it is at (x, y), and then spun in place until the required goal orientation θ is met.

It becomes clear from the course of the above explanations that the effectiveness of the proposed algorithm depends on the development of some mechanism for choosing the initial values of the pseudotemperatures (β_m) or properly adjusting the pseudotemperatures in order to generate potential fields satisfying the properties given in Section 2. Applying of an adaptive adjusting procedure to the coefficients β_m similar to that for adjusting of η_1 and η_2 (see. eq. (25)) is one possible solution. Moreover, there is no need to perform continuous adjustment throughout the whole path generation—adjustment in the first few steps gives good approximations for the initial values of the β_m coefficients. Strict solution of this problem is one of our next goals.

The proposed algorithm can be easily extended for the 3-D case (Kroumov et al. (2010)). The extension can be done by adding an additional input for the z (height of the obstacles) dimension to the obstacle description neural network. Adding of z coordinate and employing a technique for path planning, similar to that proposed in Henrich et al. (1998), would allow further adoption of the algorithm for planning of paths for industrial robots. Development of such application is another subject of our future work.

8. Acknowledgement

This work is partially supported by the Graduate School of Okayama University of Science, Okayama, Japan.

The authors would like to thank Mr. Yoshiro Kobayashi, a graduate student at the Graduate School of Okayama University of Science, for his help in preparing part of the figures. We are thankful to Ms Z. Krasteva for proofreading the final version of the manuscript.

9. References

Canny, J. F. (1988). *The Complexity of Robot Motion Planning*, The MIT Press, Cambridge, MA, U. S. A.

Chen, P. C. & Hwang, Y. K. (1998). Sandros: A dynamic graph search algorithm for motion planning, *IEEE Transactins on Robotics & Automation* 14(3): 390–403.

Faverjon, B. & Tournassoud, P. (1987). A local approach for path planning of manipulators with a high number of degrees of freedom, *Proceedings of IEEE International Conference on Robotics & Automation*, Raleigh, New Caledonia, U. S. A., pp. 1152–1159.

Ge, S. S. & Cui, Y. J. (2000). New potential functions for mobile robot path planning, *IEEE Transactins on Robotics & Automation* 16(5): 615–620.

Henrich, D., Wurll, C. & Wörn, H. (1998). 6 dof path planning in dynamic environments—a parallel on-line approach, *Proceedings of the 1998 IEEE Conference on Robotics & Automation*, Leuven, Belgium, pp. 330–335.

Hwang, Y. K. & Ahuja, K. (1992). Potential field approach to path planning, *IEEE Transactins on Robotics & Automation* Vol. 8(No. 1): 23–32.

Khatib, O. (1986). Real-time obstacle avoidance for manipulators and mobile robots, *International Journal of Robotics Research* 5(1): 90–98.

Khosla, P. & Volpe, R. (1988). Superquadratic artificial potentials for obstacle avoidance and approach, *Proceedings of IEEE International Conference on Robotics & Automation*, Leuven, Belgium, pp. 1778–1784.

Kroumov, V., Yu, J. & Negishi, H. (2004). Path planning algorithm for car-like robot and its application, *Journal of Systemics, Cybernetics and Informatics* Vol. 2(No. 3): 7.

Kroumov, V., Yu, J. & Shibayama, K. (2010). 3d path planning for mobile robots using simulated annealing neural network, *International Journal on Innovative Computing, Information and Control* 6(7): 2885–2899.

Latombe, J. C. (1991). *Robot Motion Planning*, Kluwer Academic Publishers, Norwell.

Lee, S. & Kardaras, G. (1997a). Collision-free path planning with neural networks, *Proceedings of 1997 IEEE International Conference on Robotics & Automation*, Vol. 4, pp. 3565–3570.

Lee, S. & Kardaras, G. (1997b).   Elastic string based global path planning using neural networks, *Proceedings of 1997 IEEE International Symposium on Computational Intelligence in Robotics and Automation (CIRA'97)*, pp. 108–114.

Lozano-Pèrez, T., Mason, M. T. & Taylor, R. H. (1994). Automatic synthesis of fine motion strategies for robots, *International Journal of Robotics Research* 3(1): 3–24.

Lozano-Pèrez, T. & Wesley, M. A. (1979). An algorithm for planning collision-free paths among polyhedral obstacles, *Communications of the ICM* Vol. 22(No. 10): 560–570.

Mondada, F., Franzi, E. & Ienne, P. (1993).   Mobile robot miniaturization: a tool for investigation in control algorithms, *Proceedings of ISER3*, Kyoto, Japan.

O'Dúnlaing, C. & Yap, C. K. (1982). A retraction method for planning the motion of a disk, *Journal of Algorithms* Vol. 6: 104–111.

Paden, B., Mees, A. & Fisher, M. (1989).   Path planning using a jacobian-based freespace generation algorithm, *Proceedings of IEEE International Conference on Robotics & Automation*, Scottsdale, Arizona, U. S. A., pp. 1732–1737.

Rimon, E. & Doditschek, D. E. (1992). Exact robot navigation using artificial potential fields, *IEEE Transactins on Robotics & Automation* Vol. 8(No. 5): 501–518.

Sun, Z. Q., Zhang, Z. X. & Deng, Z. D. (1997). *Intelligent Control Systems*, Tsinghua University Press.

Tsankova, D. (2010).   Neural networks based navigation and control of a mobile robot in a partially known environment, *in* A. Barrera (ed.), *Mobile Robots Navigation*, InTech, pp. 197–222.

Warren, C. W. (1989). Global path planning using artificial potential fields, *Proceedings of IEEE International Conference on Robotics & Automation*, pp. 316–321.

Yamamoto, M., Iwamura, M. & Mohri, A. (1998).   Near-time-optimal trajectory planning for mobile robots with two independently driven wheels considering dynamical constraints and obstacles, *Journal of the Robotics Society of Japan* 16(8): 95–102.

Yu, J., Kroumov, V. & Narihisa, H. (2002). Path planning algorithm for car-like robot and its application, *Chinese Quarterly Journal of Mathematics* Vol. 17(No. 3): 98–104.

Yu, J., Kroumov, V. & Negishi, H. (2003).   Optimal path planner for mobile robot in 2d environment, *Proceedings of the 7th World Multiconference on Systemics, Cybernetics and Informatics*, Vol. 3, Orlando, Florida, U. S. A., pp. 235–240.

Zelinsky, A. (1992).   A mobile robot exploration algorithm, *IEEE Transactins on Robotics & Automation* Vol. 8(No. 6): 707–717.

Zelinsky, A. & Yuta, S. (1993). Reactive planning for mobile robots using numeric potential fields, *Proceedings of Intelligent Autonomous Systems 3 (IAS3)*, pp. 84–93.

9

Reachable Sets for Simple Models of Car Motion

Andrey Fedotov[1], Valerii Patsko[1] and Varvara Turova[2]

[1]*Institute of Mathematics and Mechanics, Ural Branch of the Russian Academy of Sciences*
[2]*Mathematics Centre, Technische Universität München*
[1]*Russia*
[2]*Germany*

1. Introduction

In 1889, Andrey Andreevich Markov published a paper in "Soobscenija Charkovskogo Matematiceskogo Obscestva" where he considered four mathematical problems related to the design of railways. The simplest among these problems (and the first one in course of the presentation) is described as follows. Find a minimum length curve between two points in the plane provided that the curvature radius of the curve should not be less than a given quantity and the tangent to the curve should have a given direction at the initial point.

In 1951, Rufus Philip Isaacs submitted his first Rand Corporation Report on differential game theory where he stated and lined out a solution to the "homicidal chauffeur" problem. In that problem, a "car" with a bounded turning radius and a constant magnitude of the linear velocity tries as soon as possible to approach an avoiding the encounter "pedestrian". At the initial time, the direction of the car velocity is given.

In 1957, in American Journal of Mathematics, Lester Eli Dubins considered a problem in the plane on finding among smooth curves of bounded curvature a minimum length curve connecting two given points provided that the outgoing direction at the first point and incoming direction at the second point are specified.

Obviously, if one takes a particular case of Isaacs' problem with the immovable "pedestrian", then the "car" will minimize the length of the curve with the bounded turning radius. The arising task coincides with the problem considered by A. A. Markov. The difference from the problem by L. E. Dubins is in the absence of a specified direction at the incoming point. The fixation of incoming and outgoing directions presents in the other three problems by A. A. Markov. However, they contain additional conditions inherent to the railway construction.

In such a way the notion of a "car" which moves only forward and has bounded turning radius appeared.

In 1990, in Pacific Journal of Mathematics, James Alexander Reeds and Lawrence Alan Shepp considered an optimization problem where the object with bounded turning radius and constant magnitude of the linear velocity can instantaneously change the direction of motion to the opposite one. In a similar way, carts move around storage rooms. Thus, the model of the car that can move forward and backward has appeared.

Two models (the forward moving car with bounded turning radius; the forward and backward moving car with bounded turning radius) gave rise to numerous literature where various optimization problems related to the transportation are studied. More sophisticated models in which a moving object is considered in a more realistic way have arisen (controlled wheel, bicycle, car with two chassis, car with a trailer). Optimization problems related to such complicated tasks are extremely difficult. The simplest models serve as a "sample" showing where the situation is easy and where it becomes complex.

One of the key notion in the mathematical control theory (Pontryagin et al., 1962), (Lee & Markus, 1967), (Agrachev & Sachkov, 2004) is reachable set. The reachable set is a collection of states which can be attained within the framework of the motion model under consideration. The reachable set at given time describes a collection of states realizable at a specified time instant. The reachable set by given time is a collection of states that can be obtained on the whole time interval from an initial time instant to a specified one.

This paper is devoted to the investigation of reachable sets at given and by given time for simplest models of car motion.

2. Simple models of car motion

The simplest car dynamics are described in dimensionless variables by the following system of equations

$$\dot{x} = \sin\theta, \ \dot{y} = \cos\theta, \ \dot{\theta} = u; \quad |u| \le 1. \tag{1}$$

The variables x, y specify the center mass position in the two-dimensional plane; θ is the angle between the direction of the velocity vector and that of the vertical axis y measured clockwise from the latter. The value u acts as a control input. The control $u(t)$ specifies the instantaneous angular velocity of the vector $(\dot{x}(t), \dot{y}(t))$ of linear velocity and is bounded as $|u(t)| \le 1$.

The motion trajectories in the plane x, y are curves of bounded curvature. The paper (Markov, 1889) considers four optimization problems related to curves of bounded curvature. The first problem (Markov, 1889), p. 250, can be interpreted as a time-optimal control problem for car dynamics (1). Also, the main theorem (Dubins, 1957), p. 515, allows an interpretation in the context of time-optimal problem for such a car. In works on theoretical robotics (Laumond, 1998), an object with dynamics (1) is called Dubins' car. Model (1) is often utilized in differential game problem formulations (Isaacs, 1951; 1965).

Next in complexity is the car model by Reeds and Shepp (Reeds & Shepp, 1990):

$$\dot{x} = w\sin\theta, \ \dot{y} = w\cos\theta, \ \dot{\theta} = u; \quad |u| \le 1, |w| \le 1. \tag{2}$$

Control u changes the angular velocity, control w is responsible for instantaneous changes of the linear velocity magnitude. In particular, the car can instantaneously change its motion direction to the opposite one. The angle θ is the angle between the direction of the axis y and the direction of the forward motion of the car.

It is natural to consider control dynamics where the control w is from the interval $[a, 1]$:

$$\dot{x} = w\sin\theta, \ \dot{y} = w\cos\theta, \ \dot{\theta} = u; \quad |u| \le 1, w \in [a, 1]. \tag{3}$$

Here $a \in [-1, 1]$ is the parameter of the problem. If $a = 1$, Dubins' car is obtained; if $a = -1$, Reeds and Shepp's car appears.

Finally, one can consider non-symmetric constraint $u \in [b,1]$ instead of the bound $|u| \leq 1$. Assume that the values of the parameter b satisfy the inclusion $b \in [-1,0)$. In this case, controlled system preserves essential properties inherent to the case $|u| \leq 1$ (since the value $u = 0$ remains the internal point of the restriction on u).

Let us write down the dynamics of the last system as the most general one among the above mentioned:

$$\dot{x} = w \sin\theta, \quad \dot{y} = w \cos\theta, \quad \dot{\theta} = u; \quad u \in [b,1], \; w \in [a,1]. \tag{4}$$

Here, a and b are fixed parameters with $a \in [-1,1]$, $b \in [-1,0)$.

The model (4) is kinematic, since it does not take into account forces acting on the body. The car is represented as a point mass. The control u determines the angular rate of the linear velocity vector; the control w changes the magnitude of the linear velocity.

In the paper, reachable sets at given time and those ones by given time for system (4) are studied using numerical constructions. The reachable set at time t_* is a collection of all states which can be obtained exactly at given time t_*. If such states are considered in the plane x, y, then we have a two-dimensional reachable set; if the consideration takes place in space x, y, θ, then a three-dimensional set is obtained. The peculiarity of the reachable set "by given time" is in accounting for the states reachable not only at time t_* but on the whole interval $[0, t_*]$.

3. Two-dimensional reachable sets

3.1 Reachable sets at given and by given time

Let $z_0 = (x_0, y_0)$ be the initial geometric position and θ_0 be the initial angle at time $t_0 = 0$. The reachable set $G^2(t_*; z_0, \theta_0)$ at given time t_* in the plane x, y is the set of all geometric positions which can be reached from the starting point z_0, θ_0 at time t_* by the trajectories of system (4) using admissible piecewise continuous controls $u(\cdot), w(\cdot)$.

Introducing the notation $z(t_*; z_0, \theta_0, u(\cdot), w(\cdot))$ for the solution of differential equation (4) with the initial point z_0, θ_0, one can write

$$G^2(t_*; z_0, \theta_0) := \bigcup_{u(\cdot), w(\cdot)} z(t_*; z_0, \theta_0, u(\cdot), w(\cdot)).$$

For $z_0 = 0, \theta_0 = 0$, the notation $G^2(t_*)$ will be used instead of $G^2(t_*; 0, 0)$.

Since the right-hand side of system (4) does not contain variables x, y, and due to the type of appearance of θ in the right-hand side of (4), the geometry of reachable sets does not depend on z_0, θ_0. Namely,

$$G^2(t_*; z_0, \theta_0) = \Pi_{\theta_0}(G^2(t_*)) + z_0.$$

Here, Π_θ is the operator of clockwise rotation of the plane x, y by the angle θ_0. Therefore, the study of reachable sets for point-wise initial conditions can be restricted to the case $z_0 = 0$, $\theta_0 = 0$.

The reachable sets by given time are defined in a similar way. Let

$$\mathcal{G}^2(t_*; z_0, \theta_0) := \bigcup_{t \in [0, t_*]} G^2(t; z_0, \theta_0).$$

Other formulas remain the same, we only replace G^2 by $\mathcal{G}^2$.

Classical and the best known is the construction of the reachable sets $G^2(t_*)$ and $\mathcal{G}^2(t_*)$ for system (1). It was established in paper (Cockayne & Hall, 1975) that any point of the boundary $\partial G^2(t_*)$ can be attained using piecewise control with at most one switch. Namely, the following variants of the control actions are possible: $(-1, 0), (1, 0), (-1, 1), (1, -1)$. The first variant means that control $u \equiv 1$ acts on some interval $[0, \hat{t})$, and control $u \equiv 0$ works on the interval $[\hat{t}, t_*]$. The similar is true for the rest of the four variants. Using this proposition and running the switch instant $\hat{t}$ from 0 to t_*, we can construct the boundary of the set $G^2(t_*)$. The form of the sets $G^2(t_*)$ for four instants of time $t_* = i\,0.5\pi$, $i = 1, 2, 3, 4$, is shown in Fig. 1. The chosen values t_* correspond to the time needed to turn the velocity vector by the angle $i\,0.5\pi$ when moving with the control $u = 1$ ($u = -1$). For every t_*, the figure has its own scale. In Fig. 2, several trajectories arriving at the boundary of the reachable set are shown. The trajectories with the extreme controls $u = 1$ and $u = -1$ are the circles of radius 1. For $u = 0$, the motion along a straight line occurs.

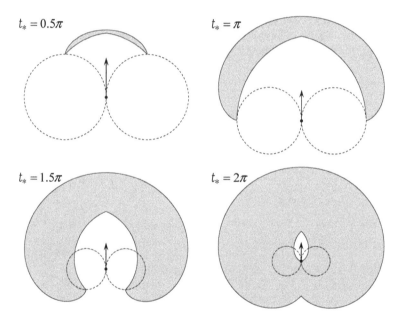

Fig. 1. Form of the reachable sets $G^2(t_*)$ for system (1)

Since the set $\mathcal{G}^2(t_*)$ is the union of the sets $G^2(t)$ over all $t \in [0, t_*]$, then any point of the boundary $\partial \mathcal{G}^2(t_*)$ is reachable with the control of the above mentioned structure (but this control is defined in general on $[0, \bar{t}]$ where $\bar{t} \in [0, t_*]$). The minimum time problem for system (1) with the initial point z_0 and given direction θ_0 of the velocity vector to the point z for which the angle θ is not specified is very popular. The structure of the solution to this problem was described by A. A. Markov. It was also known to R. Isaacs, since such problem is a degenerate case of the differential game "homicidal chauffeur". Nevertheless, accurate images of reachable sets $\mathcal{G}^2(t_*)$ have been appeared only in the beginning of 1990s (Boissonnat

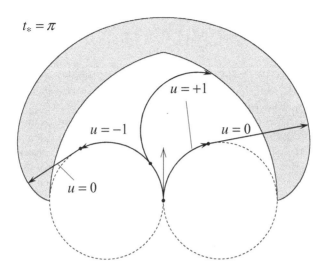

Fig. 2. Structure of the controls steering to the boundary of the reachable set $G^2(\pi)$ for system (1)

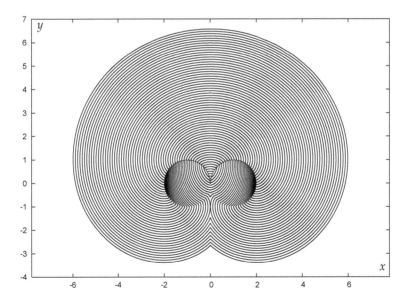

Fig. 3. Time-limited reachable sets $\mathcal{G}^2(t)$ for system (1), $\tau_f = 7.3$, $\delta = 0.1$

& Bui, 1994). Fig. 3 shows the form of the sets $\mathcal{G}^2(t)$ computed by the authors on the interval $[0, t_f]$, $t_f = 7.3$, the output step for the construction results is $\delta = 0.1$.

3.2 Reachable sets with regard to an orientedly added set

Let us now define the reachable set at time t_* with regard to an orientedly added set D:

$$G_D^2(t_*) := \bigcup_{u(\cdot),w(\cdot)} [z(t_*;0,0,u(\cdot),w(\cdot)) + \Pi_{\theta(t_*;0,0,u(\cdot),w(\cdot))}(D)].$$

Hence, when constructing $G_D^2(t_*)$, we add the set D which is rigidly rotated with respect to the origin by the angle $\theta(t_*;0,0,u(\cdot),w(\cdot))$ to each point (being attainable at the time t_* with $u(\cdot),w(\cdot)$) of the usual reachable set. It is assumed that $z_0 = 0$ and $\theta_0 = 0$.

The sense of the set $G_D^2(t_*)$ can be explained using the following example. Let us fix controls $u(\cdot),w(\cdot)$. At the time t_*, the geometric position $z(t_*;0,0,u(\cdot),w(\cdot))$ and the slope $\theta(t_*;0,0,u(\cdot),w(\cdot))$ of the velocity vector are realized. Suppose we are interested at this time instant in a point located at the distance d from the point $z(t_*;0,0,u(\cdot),w(\cdot))$ orthogonally to the velocity vector $\dot{z}(t_*) = (\dot{x}(t_*),\dot{y}(t_*))$ to the left from its direction. Such point is written as $z(t_*;0,0,u(\cdot),w(\cdot)) + \Pi_{\theta(t_*;0,0,u(\cdot),w(\cdot))}(D)$, if we take the set consisting from the point $(-d,0)$ in the plane x,y as the set D. The total collection of points at the time t_* with the property we are interested in is obtained by the enumeration of admissible controls $u(\cdot)$, $w(\cdot)$ and forms the set $G_D^2(t_*)$.

The reachable set $\mathcal{G}_D^2(t_*)$ by the time t_* with regard to an orientedly added set D is defined as

$$\mathcal{G}_D^2(t_*) := \bigcup_{t \in [0,t_*]} G_D^2(t).$$

3.3 Isaacs' transformation

System (4) is of the third order with respect to the state variables. Since we are interested in the construction of reachable sets in the plane of geometric coordinates, it is convenient to pass to a system of the second order with respect to the state variables. This can be done using Isaacs' transformation.

Isaacs utilized system (4) with $w \equiv 1$, $u \in [-1,1]$ (i.e. system (1)) for the formulation and solution of several pursuit-evasion differential games. The pursuit-evasion game "homicidal chauffeur" proposed by R. Isaacs in his report (Isaacs, 1951) was then published in his famous book "Differential games" (Isaacs, 1965) and became classical problem. We will apply the transformation, which R. Isaacs used in this game, for our purposes.

Let $h(t)$ be a unit vector in the direction of forward motion of system (4) at time t. The orthogonal to $h(t)$ unit vector is denoted by $k(t)$ (see Fig. 4). We have

$$h(t) = \begin{pmatrix} \sin \theta(t) \\ \cos \theta(t) \end{pmatrix}, \quad k(t) = \begin{pmatrix} \cos \theta(t) \\ -\sin \theta(t) \end{pmatrix}.$$

Axis $\mathbf{y}$ of the reference system is directed along the vector $h(t)$, axis $\mathbf{x}$ is directed along the vector $k(t)$.

Let $\tilde{z} = (\tilde{x},\tilde{y})$ be a fixed point in the plane x,y. The coordinates $\mathbf{x}(t),\mathbf{y}(t)$ of this point in the movable reference system whose origin coincides with the point $z(t)$ are

$$\mathbf{x}(t) = k'(t)(\tilde{z} - z(t)) = \cos \theta(t)(\tilde{x} - x(t)) - \sin \theta(t)(\tilde{y} - y(t)), \tag{5}$$

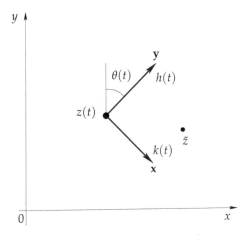

Fig. 4. Movable reference system

$$\mathbf{y}(t) = h'(t)(\tilde{z} - z(t)) = \sin\theta(t)(\tilde{x} - x(t)) + \cos\theta(t)(\tilde{y} - y(t)). \tag{6}$$

Here, the prime denotes the operation of transposition.
Taking into account (5) and (6), one obtains

$$\dot{x}(t) = -\sin\theta(t)\dot{\theta}(t)(\tilde{x} - x(t)) - \cos\theta(t)\dot{x}(t) -$$

$$\cos\theta(t)\dot{\theta}(t)(\tilde{y} - y(t)) + \sin\theta(t)\dot{y}(t) = -\mathbf{y}(t)\dot{\theta}(t) - \cos\theta(t)\dot{x}(t) + \sin\theta(t)\dot{y}(t), \tag{7}$$

$$\dot{y}(t) = \cos\theta(t)\dot{\theta}(t)(\tilde{x} - x(t)) - \sin\theta(t)\dot{x}(t) -$$

$$\sin\theta(t)\dot{\theta}(t)(\tilde{y} - y(t)) - \cos\theta(t)\dot{y}(t) = \mathbf{x}(t)\dot{\theta}(t) - \sin\theta(t)\dot{x}(t) - \cos\theta(t)\dot{y}(t). \tag{8}$$

The projection of the velocity vector of system (4) on the axis **x** at time t is zero, the projection on the axis **y** is $w(t)$. Therefore,

$$\cos\theta(t)\dot{x}(t) - \sin\theta(t)\dot{y}(t) = 0, \tag{9}$$

$$\sin\theta(t)\dot{x}(t) + \cos\theta(t)\dot{y}(t) = w(t). \tag{10}$$

Using (7) and (9), we obtain

$$\dot{x}(t) = -\mathbf{y}(t)\dot{\theta}(t). \tag{11}$$

With (8) and (10), we get

$$\dot{y}(t) = \mathbf{x}(t)\dot{\theta}(t) - w(t). \tag{12}$$

Equalities (11), (12), and $\dot{\theta} = u$ yield the system

$$\dot{x} = -yu,$$
$$\dot{y} = xu - w, \quad u \in [b, 1], \quad w \in [a, 1]. \tag{13}$$

3.4 Solvability sets for problems of approach at given time and by given time

Let M be a closed set in the plane x, y. Denote by $W(\tau, M)$ (respectively, by $\mathcal{W}(\tau, M)$) the set of all points $z = (x, y)$ with the property: there exist piecewise continuous admissible controls $u(\cdot), w(\cdot)$ which bring system (13) from the initial point z to the set M at time τ (respectively, by time τ). The set $W(\tau, M)$ ($\mathcal{W}(\tau, M)$) is the solvability set in the problem of reaching the set M at time τ (by time τ).

Take now the same set M as a terminal set in the problem of reaching a given set for system (13) and as a set D in the problem of finding reachable set in geometric coordinates for system (4) with regard to an orientedly added set. It follows from the sense of Isaacs' transformation that the set $W(\tau, M)$ drawn in coordinates x, y coincides with the set $G_M^2(\tau)$ depicted in coordinates x, y. Therefore,

$$W(\tau, M) = G_M^2(\tau). \tag{14}$$

Similarly,

$$\mathcal{W}(\tau, M) = \mathcal{G}_M^2(\tau). \tag{15}$$

In the following, we will utilize relations (14), (15) in order to obtain the reachable sets $G_D^2(t)$ and $\mathcal{G}_D^2(t)$ of system (4) using the sets $W(\tau, M)$ and $\mathcal{W}(\tau, M)$ computed for system (13) with $\tau = t$ and $M = D$. In addition, if M is a one-point set that coincides with the origin in the plane x, y, then

$$W(\tau, M) = G_M^2(\tau) = G^2(\tau), \quad \mathcal{W}(\tau, M) = \mathcal{G}_M^2(\tau) = \mathcal{G}^2(\tau).$$

By fixing some point $\bar{z} = (\bar{x}, \bar{y})$ in the plane and by increasing t, let us find the first instant $\bar{t}$ when $\bar{z} \in \mathcal{G}_M^2(\bar{t})$ (equivalently, $\bar{z} \in G_M^2(\bar{t})$). Such $\bar{t}$ be the optimal time $V(\bar{z})$ of passing from the point $z_0 = 0, \theta_0 = 0$ to the point $\bar{z}$ for system (4) with accounting for M. Hence, the Lebesgue set $\{z : V(z) \leq t\}$ of the optimal result function coincides with the set $\mathcal{G}_M^2(t) = \mathcal{W}(t, M)$. For the sets shown in Fig. 3 (where $M = \{0\}$), the function $z \to V(z)$ is discontinuous on the upper semi-circumferences of radius 1 with the centers at the points $(-1, 0)$, $(1, 0)$.

3.5 Backward procedure for construction of solvability sets

The algorithms developed by the authors for the numerical construction of the sets $W(\tau, M)$ and $\mathcal{W}(\tau, M)$ are based on the backward procedure (the parameter τ increases starting from $\tau = 0$) and being variants of the dynamic programming method for the considered class of problems. Backward procedures for the construction of the solvability sets at given time and by given time are intensively developed (Grigor'eva et al., 2005), (Mikhalev & Ushakov, 2007) for control problems and differential games. Elements of the backward constructions are included in one or another form (Sethian, 1999), (Cristiani & Falcone, 2009) into grid methods for solving differential games. For control problems, the backward procedures are simpler, since the second player whose actions should be accounted for in differential games is absent. Especially simple are the backward procedures for problems in the plane. In this case, one can storage the boundary of the sets $W(\tau, M)$ and $\mathcal{W}(\tau, M)$ in form of polygonal lines.

The general idea for the construction of the solvability sets $W(\tau, M)$ in the problem of approaching a set M by system (13) at time τ is the following. We deal with polygonal sets which are considered as approximations of the ideal sets $W(\tau, M)$. Specify a time step Δ (it can also be variable) and define $\tau_0 = 0, ..., \tau_{i+1} = \tau_i + \Delta$.

Consider the vectogram $E(\mathbf{z}) = \bigcup\limits_{\mathbf{u} \in P} f(\mathbf{z}, \mathbf{u})$ of the right hand side of the controlled system.

For system (4), we have $\mathbf{u} = (u, w)$, $P = \{(u, w) : u \in [b, 1], w \in [a, 1]\}$. The set $\mathbf{z} - \Delta E(\mathbf{z})$ describes approximately the collection of points from which the point $\mathbf{z}$ can be approached at time Δ. Running over the boundary $\partial W(\tau_i, M)$ and attaching the set $\mathbf{z} - \Delta E(\mathbf{z})$ "to every" point of the boundary, we approximately construct the boundary $\partial W(\tau_{i+1}, M)$ of the set $W(\tau_{i+1}, M)$.

Theoretically, the solvability set $\mathcal{W}(\tau_*, M)$ for the problem of approaching the set M by system (13) by the time τ_* is defined through the union of the sets $W(\tau, M), \tau \in [0, \tau_*]$. However, one can reject the explicit realization of the union operation and try to construct the boundary of the set $\mathcal{W}(\tau_{i+1}, M)$ directly on the base of the boundary of the set $\mathcal{W}(\tau_i, M)$. To this end, the notion of a front is introduced.

Using the optimal result function V, the front corresponding to the time τ is formally defined as

$$F(\tau) := \{\mathbf{z} \in \partial W(\tau, M) : V(\mathbf{z}) = \tau\}.$$

If $\bar{\mathbf{z}} \notin W(\tau, M)$, then every trajectory of system (13) starting from the point $\bar{\mathbf{z}}$ can approach $W(\tau, M)$ through the front $F(\tau)$ only.

It is known (see e.g. Bardi & Capuzzo-Dolcetta (1997)) that if $F(\tau_*) = \partial W(\tau_*, M)$ for some τ_*, then $F(\tau) = \partial W(\tau, M)$ for all $\tau \geq \tau_*$.

Let $A(\tau) := \partial W(\tau, M) \setminus F(\tau)$. It follows from results of optimal control theory and theory of differential games that the function $\mathbf{z} \to V(\mathbf{z})$ is discontinuous on the set $\mathcal{A} := \bigcup\limits_{\tau \geq 0} A(\tau)$ and continuous in the remaining part of the plane.

Possible structure of the set $\mathcal{A}$ is not well explored for time-optimal problems. It is reasonable to assume that if the set $\mathcal{B} := \mathcal{A} \setminus M$ is not empty, then it consists of a collection of smooth arcs. Such arcs are called barriers (Isaacs, 1965).

By the definition of the front $F(\tau)$, we have $F(\tau) \subset W(\tau, M)$. From here, with accounting for the relations $F(\tau) \subset \partial W(\tau, M), W(\tau, M) \subset \mathcal{W}(\tau, M)$, we obtain $F(\tau) \subset \partial W(\tau, M)$.

The main idea of the algorithm of the backward construction of the sets $W(\tau, M)$ is explained in Fig. 5. The next set $W(\tau_{i+1}, M)$ for $\tau_{i+1} = \tau_i + \Delta$ is computed on the basis of the previous set $W(\tau_i, M)$. The central role in this computation belongs to the front $F(\tau_i)$. As a result, the front $F(\tau_{i+1})$ is obtained, and a new set $A(\tau_{i+1})$ is formed via the extension or reduction of the set $A(\tau_i)$. The union $F(\tau_{i+1}) \cup A(\tau_{i+1})$ is the boundary of the next set $W(\tau_{i+1}, M)$. The initial front $F(0)$ coincides with those part of the boundary ∂M that consists of the points from which the trajectories of system (13) (being written in backward time) leave M with increasing τ for at least one pair of admissible controls $u(\cdot), w(\cdot)$. According to (Isaacs, 1965), such part of the boundary of M is called the usable part of M.

Therefore, the algorithm resembles the front propagation.

Several typical cases (local fragments) of the front propagation are presented in Figs. 6 and 7. Fig. 6a shows the case in which the left end of the front is moving from the endpoint c of the usable part of ∂M with increasing τ. In the algorithm, simultaneously with the computation of the next front $F(\tau_{i+1})$, the extension of the barrier is computed by means of connection of the left ends of the fronts $F(\tau_i)$ and $F(\tau_{i+1})$. In the case considered, this results in the local increase of the totality $A(\tau_{i+1})$ with respect to $A(\tau_i)$. The extension of the barrier forms a line on which the value function is discontinuous.

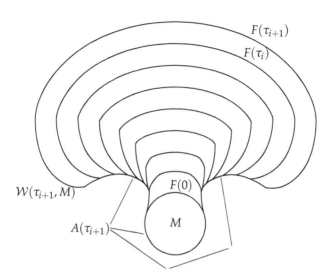

Fig. 5. Backward construction of the solvability sets $W(\tau, M)$. The boundary of the set $W(\tau_{i+1}, M)$ is $F(\tau_{i+1}) \cup A(\tau_{i+1})$

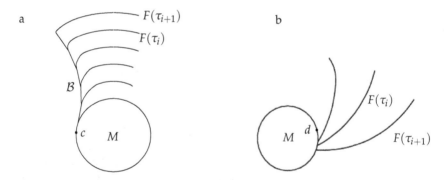

Fig. 6. a) The movement of the left end of the front generates the barrier line on which the value function is discontinuous; b) The right end of the front is moving along the boundary of the terminal set

In the case shown in Fig. 6b, the right end of the front starts to move along ∂M from the very beginning i.e. for small $\tau > 0$. Here, no barrier line emanates from the right endpoint d of the usable part. The value function near the point d outside the set M is continuous.

Fig. 7 represents the case where the left end of the front is running along the back side of the already constructed barrier. This results in the local decrease of the totality $A(\tau_{i+1})$ comparing to $A(\tau_i)$.

A more detailed description of the algorithm is given in (Patsko & Turova, 2009).

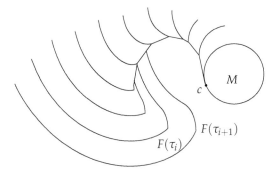

Fig. 7. The left end of the front is bending around the barrier line

3.6 Results of numerical construction of solvability sets (reachable sets)

Let M be a one-point set that coincides with the origin (in the numerical computations, a circle with a very small radius is taken). For system (3) with $a < 0$, the set $G^2(t)$ "swells" monotonically with increasing t, i.e. $G^2(t_2) \supset G^2(t_1)$ for $t_2 > t_1$, where the strict inclusion holds. This provides that the sets $G^2(t)$ and $\mathcal{G}^2(t)$ coincide. For $a = -1$ (i.e. for system (2)) the set $G^2(t) = \mathcal{G}^2(t)$ is symmetric with respect to the axes x, y.

After publishing the paper (Reeds & Shepp, 1990) related to the minimum time problem for system (2), the obtained results were refined and essentially propelled in the works (Sussmann & Tang, 1991), (Soueres et al., 1994), and (Soueres & Laumond, 1996) by using the Pontryagin maximum principle. The second paper describes in particular the construction of reachable sets $G^2(t)$ and give the structure of controls steering to the boundary of these sets. The properties of monotonic swelling of the sets $G^2(t)$ and the symmetry make system (2) very convenient for solving very complex problems of robot transportation (Laumond, 1998), pp. 23–43. For $a = -0.8$ and $a = -1$, the reachable sets are shown in Fig. 8. As before, the notation t_f means the end time of the construction interval. The symbol δ denotes the output step of the representation, which is not necessarily equal to the step Δ of the backward constructions. The latter is, as a rule, smaller.

For $a = 0.8$ and $a = 0.2$, the behavior of the reachable sets $G^2(t)$ for system (3) with increasing t is shown in Fig. 9. The dependency of the sets $G^2(t)$ and $\mathcal{G}^2(t)$ on the parameter a is presented for $t = 1.8$ in Fig. 10. Similar sets but for non-symmetric constraint $u \in [-0.6, 1]$ (i.e. for system (4)) are depicted in Fig. 11. Non-symmetry of the restriction on the control u results in the non-symmetry of the obtained sets with respect to the vertical axis. Note that from the theoretical point of view, the minimum time problem and the related problem of the construction of the sets $\mathcal{G}^2(t)$ for system (4) with $a = 1$ and $b \in [-1, 0)$ were studied in (Bakolas & Tsiotras, 2011).

In Figs. 12 and 13, the sets $\mathcal{G}^2_M(\tau)$ computed for $a = 0.2$ and $b = -0.6$ are shown. The set $\mathcal{G}^2_M(\tau) = W(\tau, M)$ becomes for the first time non-simply-connected at $\tau_* = 4.452$ when the right part of the front collides with M. Here, one needs to fill a "hole" adjoining to the back side of initial part of the barrier. The hole is completely filled out by the time $\bar{\tau} = 4.522$. The second hole occurs at time $\tau^* = 5.062$ when left and right parts of the front meet. In this case, the filling out of the hole ends at $\bar{\bar{\tau}} = 5.21$. The function V of the optimal result is

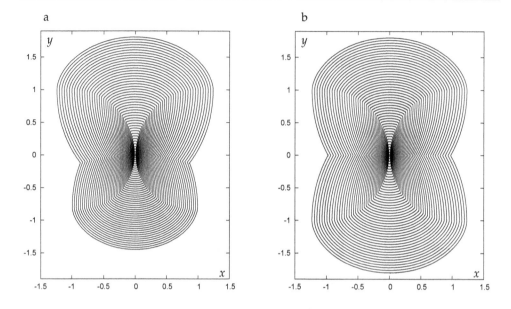

Fig. 8. Reachable sets $\mathcal{G}^2(t) = G^2(t)$ for system (3), $t_f = 1.8$, $\delta = 0.04$: a) $a = -0.8$; b) $a = -1$

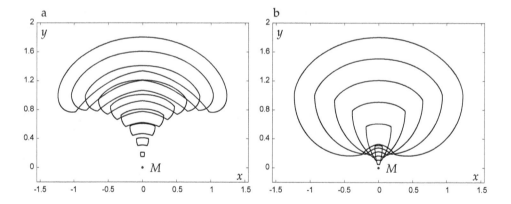

Fig. 9. The reachable sets $G^2(t)$ at given time for system (3), $t_f = 1.8$: a) $a = 0.8$, $\delta = 0.2$; b) $a = 0.2$, $\delta = 0.3$

discontinuous on the two barrier lines being the upper semi-circumferences with the centers at the points $(a/b, 0) = (-1/3, 0)$ and $(a, 0) = (0.2, 0)$ and the radiuses $\frac{1}{3} - r$ and $0.2 - r$, where $r = 0.01$ is the radius of the terminal circle M.

Let us consider an example where the set M is a circle of radius 0.3 centered at the point $(0.7, 0)$. Put $a = 0.8$. In Figs. 14 and 15, results of the construction of the sets $\mathcal{W}(\tau, M)$ are presented. We see which parts of the boundary of the reachable set $G_M^2(\tau) = W(\tau, M)$

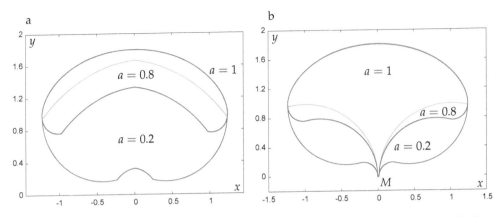

Fig. 10. Comparison of the reachable sets for system (3) for different values of a, $t = 1.8$: a) $G^2(t)$; b) $\mathcal{G}^2(t)$

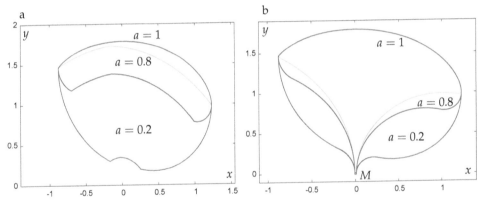

Fig. 11. Comparison of the reachable sets for system (4) for different values of a, $t = 1.8$: a) $G^2(t)$; b) $\mathcal{G}^2(t)$

propagate in a regular way with increasing τ and which ones (and from what time) are developed in a more complex manner. With increasing τ, the left end of the front moves along the barrier line (as in Fig. 6a). After passing the point d, the left end begins to move along the same barrier line but over its back side. The right end runs along the boundary of the terminal set with increasing τ (as in Fig. 6b), then changes over to the back side of the barrier. At time $\tau^* = 4.864$, the self-intersection of the front occurs. The front is divided into two parts: the inner and the outer ones. The ends of the inner front slide along the left barrier until they meet each other and a closed front is formed. The construction then is continued for the closed front. At time $\tau^* = 6.08$, the inner front shrinks into a point. The computation of the closed outer front is also continued till this time τ^*. The optimal result function V is discontinuous outside of M on the barrier ce.

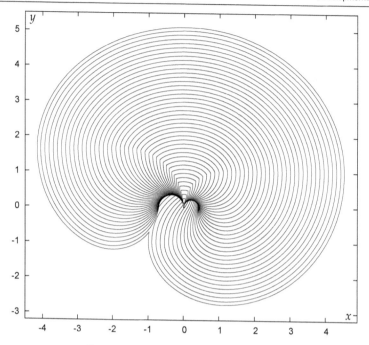

Fig. 12. Reachable sets $\mathcal{G}_M^2(\tau) = \mathcal{W}(\tau, M)$ for system (4) with $a = 0.2$, $b = -0.6$. The set M is the circle of radius 0.01 with the center at the origin. The output step of the representation is $\delta = 0.12$, the terminal time is $\tau_f = 5.21$

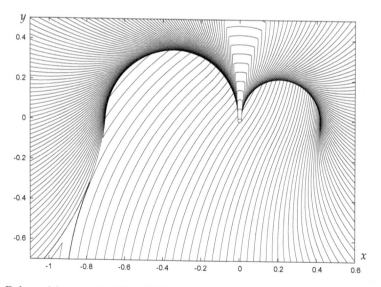

Fig. 13. Enlarged fragment of Fig. 12. The output step of the representation is $\delta = 0.046$

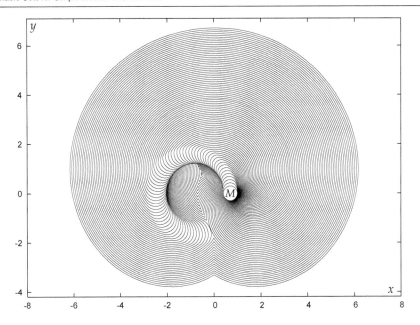

Fig. 14. Reachable sets $\mathcal{G}_M^2(\tau) = \mathcal{W}(\tau, M)$ for the circle M centered at $(0.7, 0)$; $a = 0.8$, $b = -1$, $\tau_f = 6.08$, $\delta = 0.076$

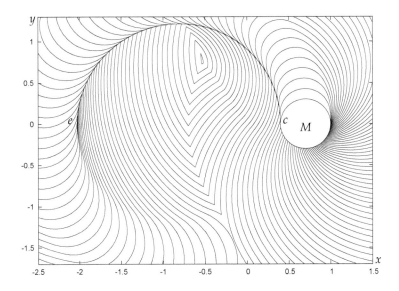

Fig. 15. Enlarged fragment of Fig. 14

4. Three-dimensional reachable sets

Let us describe the reachable sets

$$G^3(t_*) := \bigcup_{u(\cdot), w(\cdot)} (z(t_*; 0, 0, u(\cdot), w(\cdot)), \theta(t_*; 0, 0, u(\cdot), w(\cdot)), \quad \mathcal{G}^3(t_*) := \bigcup_{t \in [0, t_*]} G^3(t)$$

in the three-dimensional space x, y, θ. We restrict ourselves to the case of system (1) and a close to it system in which the control parameter u is restricted as $u \in [b, 1]$, where $b \in [-1, 0)$ is the fixed parameter.

4.1 Structure of controls steering to the boundary of reachable sets at given time

In paper (Patsko et al., 2003), it was established based on the application of the Pontryagin maximum principle to system (1) that for any point $(z, \theta) \in \partial G^3(t_*)$ the control steering to this point is piecewise continuous and has at most two switches. In addition, there are only 6 variants of changing the control:

1) 1, 0, 1;   2) -1, 0, 1;   3) 1, 0, -1;   4) -1, 0, -1;   5) 1, -1, 1;   6) -1, 1, -1.

The second variant means that the control $u \equiv -1$ acts on some interval $[0, t_1)$, the control $u \equiv 0$ works on an interval $[t_1, t_2)$, and the control $u \equiv 1$ operates on the interval $[t_2, t_*]$. If $t_1 = t_2$, then the second interval (where $u \equiv 0$) vanishes, and we obtain a single switch from $u = -1$ to $u = 1$. In the case $t_1 = 0$, the first interval where $u \equiv -1$ vanishes; in the case $t_2 = t_*$ the third interval with $u \equiv 1$ is absent. The control has constant value for all $t \in [0, t_*]$ if one of the following three conditions holds: $t_1 = t_*$, $t_2 = 0$, or both $t_1 = 0$ and $t_2 = t_*$. Similar is true for the other variants.

The proposition on six variants of the control $u(t)$ steering to the boundary of the reachable set $G^3(t_*)$ is similar in form to the Dubins theorem on the variants of the controls steering to the boundary of the reachable set $\mathcal{G}^3(t_*)$. The same variants are valid. However, due to the relation between the sets $G^3(t_*)$ and $\mathcal{G}^3(t_*)$ (the set $\mathcal{G}^3(t_*)$ is the union of the sets $G^3(t)$ over $t \in [0, t_*]$), the above mentioned properties of the controls leading to the boundary of the set $G^3(t_*)$ result in the analogous properties of the controls leading to the boundary of the set $\mathcal{G}^3(t_*)$, but the converse is false.

4.2 Numerical construction of three-dimensional reachable sets at given time

Let us apply the above formulated result on the structure of the control $u(t)$ steering to $\partial G^3(t_*)$ for the numerical construction of the boundary $\partial G^3(t_*)$.

To construct the boundary $\partial G^3(t_*)$ of the set $G^3(t_*)$, we search through all controls of the form 1–6 with two switches t_1, t_2. For every variant of switches, the parameter t_1 is chosen from the interval $[0, t_*]$, and the parameter t_2 from the interval $[t_1, t_*]$. In addition, controls with one switch and without switches are also considered. Taken a specific variant of switching and searching through the parameters t_1, t_2 on some sufficiently fine grid, we obtain a collection of points generating a surface in the three-dimensional space x, y, θ.

Therefore, each of the six variants yields its own surface in the three-dimensional space. The boundary of the reachable set $G^3(t_*)$ is composed of pieces of these surfaces. The six surfaces are loaded into the visualization program without any additional processing of data. Using this program, the boundary of the reachable sets is extracted. Some surfaces (in part or as a whole) find themselves inside of the reachable set. The visualization program does not plot such pieces.

The visualization of the three-dimensional sets is done with the program "Cortona VRML Client" utilizing the open standard format VRML/X3D for the demonstration of interactive vector graphics.

Fig. 16 shows the boundary of the set $G^3(t_*)$ at time $t_* = 1.5\pi$ from two perspectives. The initial values of x_0, y_0, and θ_0 are equal to zero. The different parts of the boundary are marked with different colors. For example, part 2 is reachable for the trajectories with the control $u(t)$ of the form $-1, 0, 1$ with two switches. The sections of the reachable set by the plane $\theta = \text{const}$ are depicted with some step along the axis θ. The points of junction lines of parts 1,2; 1,3; 2,4; 2,5; 2,6; 3,4; 3,5; 3,6 are obtained with a single-switch control. Any point of the common line of parts 5 and 6 is reachable for two trajectories with two switches each. Parts 5 and 6 have non-smooth junction along this line. The angle of the junction is not visible because it is rather small. The control $u(t) \equiv 0$ steers to the junction point of parts $1-4$.

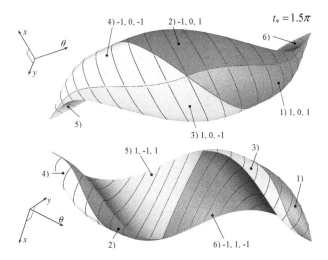

Fig. 16. The set $G^3(t_*)$ for $t_* = 1.5\pi$ shown from the two perspectives

Fig. 17 shows reachable sets $G^3(t_*)$ at the same perspective but with different scales for four time instants t_*. The transformation of the structure of the reachable set boundary is clearly seen. With increasing time, the forward part of the boundary covers the back part composed of patches 5, 6. Note that the angle θ is not restricted as $\theta \in [-\pi, \pi)$.

Passing from $t_* = 3\pi$ to $t_* = 4\pi$, one arrives at the time $t_* \approx 3.65\pi$ when the reachable set $G^3(t_*)$ becomes non-simply-connected for some small time interval. Namely, a cavity that does not belong to the reachable set arises. In Fig. 18, an origination of such a situation is shown. Here, a cut of two sets $G^3(t_*)$ corresponding to instants $t_* = 3\pi$ and $t_* = 3.65\pi$ is depicted. The cut is done using the plane $\theta = 0$. The set $G^3(3\pi)$ is simply connected and the set $G(3.65\pi)$ is not.

Fig. 19 shows the set $G^3(t_*)$ for $t_* = 1.6\pi$, $t_* = 2\pi$, and $t_* = 2.5\pi$. Here the angle θ is calculated modulo 2π.

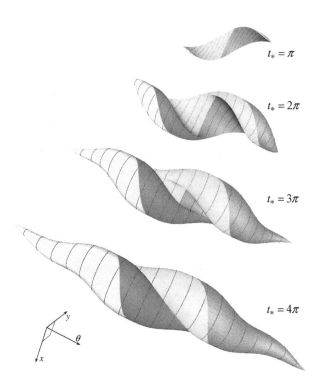

Fig. 17. Development of the reachable set $G^3(t_*)$

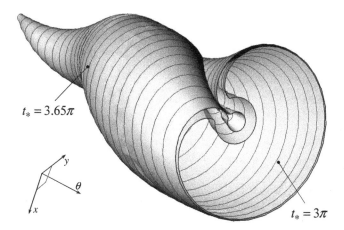

Fig. 18. Loss of simple connectivity

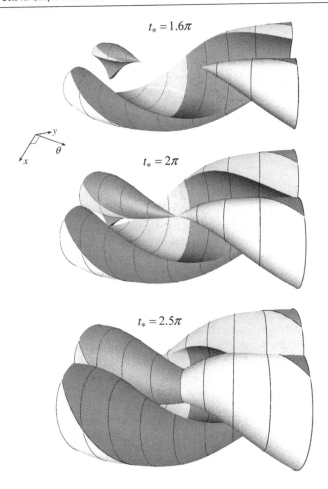

Fig. 19. The set $G^3(t_*)$ for three time instants with θ computed modulo 2π

4.3 Numerical construction of three-dimensional reachable sets by given time

Let us describe the reachable sets $\mathcal{G}^3(t_*)$ by given time. Theoretically, their construction can base on the definition $\mathcal{G}^3(t_*) = \bigcup\limits_{t \in [0,t_*]} G^3(t)$, and the boundary of the sets $G^3(t)$ can be obtained by running t on $[0, t_*]$ with a small step. However, this is very difficult approach for practical constructions. The analysis of the development of the sets $G^2(t_*)$ and $\mathcal{G}^2(t_*)$ in the plane x, y suggests a more thrifty method.

The observation of change of the set $G^3(t)$ gives the following.

For $t_* \in (0, \pi)$, any point inside of that part of the boundary $\partial G^3(t_*)$ which is generated by the controls of the kind 1–4 is strictly inside of the set $G^3(t_* + \Delta t)$ for any sufficiently small $\Delta t > 0$. Conversely, any point lying inside of that part of the boundary $\partial G^3(t_*)$ which is generated by the controls of the kind 5,6 is outside of the set $G^3(t_* + \Delta t)$.

Therefore, one can say that for $t_* \in (0, \pi)$, the front part of the boundary $\partial G^3(t_*)$ be the piece of the boundary generated by the controls 1–4, and the back part be the piece of the boundary $\partial G^3(t_*)$ corresponding to the controls 5,6. The junction of the front and back parts of the boundary $\partial G^3(t_*)$ occurs along two one-dimensional arcs in space x, y, θ, which correspond to the controls of the form (1,-1) and (-1,1) with one switch on $[0, t_*]$. Computing the collection of such arcs for every $t \in [0, t_*]$, we obtain a surface which forms the "barrier" part of the boundary of the set $\mathcal{G}^3(t_*)$. In total, the boundary of the set $\mathcal{G}^3(t_*)$ is composed of the front and barrier parts.

Thus, the construction of the boundary $\partial \mathcal{G}^3(t_*)$ for $t_* \in [0, \pi]$ requires loading of 4 surfaces corresponding to the controls 1–4 with two switches and 2 surfaces corresponding to the controls of the form (1,-1) (surface I) and (-1,1) (surface II) with one switch on the interval $[0, t]$, where $t \in [0, t_*]$, to the visualization program. The program constructs automatically the visible from the outside boundary of the set $\mathcal{G}^3(t_*)$.

Let now $t_* \in (\pi, 4\pi]$. In this case, some part of the boundary $\partial G^3(t_*)$ generated by the controls of the form 5 and 6 becomes the front one. For the construction of the boundary $\partial \mathcal{G}^3(t_*)$, 6 surfaces corresponding to the controls 1–6 with two switches and 2 surfaces I and II corresponding to the controls of the form (1,-1) and (-1,1) with one switch on $[0, t]$, where $t \in [0, t_*]$, are loaded into the visualization program. Note that for $t_* \in [2\pi, 4\pi]$ it is not necessary to increment two latter surfaces. It is sufficient to use their parts constructed up to time 2π. It should be emphasized that similarly to the case of the sets $G^3(t_*)$, there is a small time interval from $[3\pi, 4\pi]$ on which the set $\mathcal{G}^3(t_*)$ becomes non-simply-connected. For t_* from such an interval, the above described rule of the construction of the boundary using the visualization program gives only the external part of the boundary of the sets $\mathcal{G}^3(t_*)$. The detection of the "internal" boundary requires additional analysis and is not described here. Starting from the instant of time $t_* = 4\pi$, the boundary of the set $G^3(t_*)$ becomes entirely a front. In this case, $\mathcal{G}^3(t_*) = G^3(t_*)$, $t_* \geq 4\pi$.

The set $\mathcal{G}^3(t_*)$ for $t_* = 1.5\pi$ is shown from two perspectives in Fig. 20; development of $\mathcal{G}^3(t_*)$ with increasing t_* is given in Fig. 21. These pictures can be compared with Figs. 16 and 17. The difference of the reachable sets by given time from the reachable sets at given time is in the presence of the barrier part formed by the smooth surfaces I and II. To understand better its arrangement, Fig. 22 gives the cut-off sets $\mathcal{G}^3(2\pi)$ and $\mathcal{G}^3(3\pi)$ (cutting plane is $\theta = 0$). The barrier part is developed from the initial point $(0,0,0)$ (white point in the pictures). Every of the shown level lines on the barrier part corresponds to its own instant of time. Till $t_* = \pi$, the level lines are closed curves. With t_* increasing, new level lines which are not anymore closed occur. In addition, the old level lines are reduced at their ends, among them those ones constructed until $t_* = \pi$. Starting from the instant $t_* = 2\pi$, the reduction of the constructed lines begins. It finishes at the time $t_* = 4\pi$ when the set $G^3(4\pi)$ captures the point $(0,0,0)$. The set $\mathcal{G}^3(t_*)$ for $t_* = 2\pi$ with the angle θ computed modulo 2π is shown in Fig. 23.

Time-dependent construction of the reachable sets $\mathcal{G}^3(t_*)$ with the indication of which control from variants 1–6 corresponds to every particular piece on the front part of these sets is close to finding of optimal feedback control synthesis for the minimum time problem of steering to the point $(0,0,0)$. The optimal feedback control synthesis for system (1) was obtained in (Pecsvaradi, 1972) (θ is taken modulo 2π).

Note that some individual images of three-dimensional reachable sets by given time (with θ taken modulo 2π) obtained by other means are available in the works (Laumond, 1998), p. 7, and (Takei & Tsai, 2010), pp. 22, 23, 26, and 27.

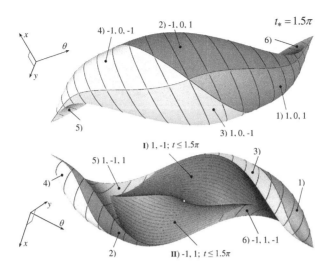

Fig. 20. The set $\mathcal{G}^3(t_*)$ for $t_* = 1.5\pi$ shown from the two perspectives

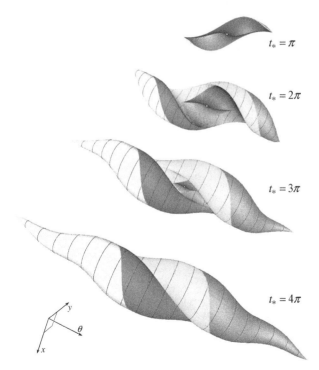

Fig. 21. Development of the reachable set $\mathcal{G}^3(t_*)$

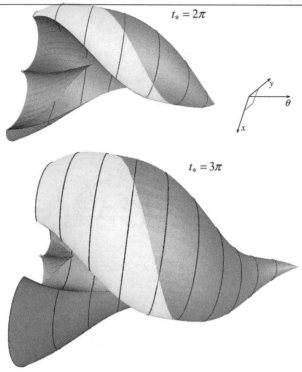

Fig. 22. The cut-off sets $\mathcal{G}^3(t_*)$ for $t_* = 2\pi$ and $t_* = 3\pi$, cutting plane $\theta = 0$

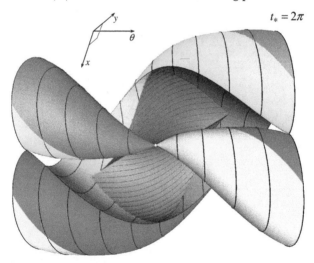

Fig. 23. The set $\mathcal{G}^3(2\pi)$ with the angle θ computed modulo 2π

4.4 Case of non-symmetric constraint on control u

The proposition on the structure of controls steering trajectories of system (1) to the boundary of the reachable set $G^3(t_*)$ is also preserved for the case of non-symmetric constraint $u \in [b, 1]$ with $b \in [-1, 0)$. One should only replace $u = -1$ by $u = b$. Results of the construction of the sets $G^3(t_*)$ are shown for $b = -0.25$ in Fig. 24. The sets $G^3(4\pi)$ and $G^3(6\pi)$ are depicted from the same perspective and have the same scale. Approximately the same perspective but a larger scale is used in Fig. 25 presenting the set $\mathcal{G}^3(4\pi)$. This set with the angle θ taken modulo 2π is shown in Fig. 26.

With a fixed point (x, y, θ), we can compute the first instant $V(x, y, \theta)$ when this point is on the boundary of the set $G^3(t)$ or, what is the same, on the boundary of $\mathcal{G}^3(t)$. The value $V(x, y, \theta)$ be the optimal steering time from the point $(0, 0, 0)$ to the point (x, y, θ). Paper (Bakolas & Tsiotras, 2011) gives results on the computation of level sets of the function $V(x, y, \theta)$ for fixed values θ (modulo 2π) and different values of the parameter b. This is equivalent to the computation of θ-sections of the sets $\mathcal{G}^3(t)$ for different values of b.

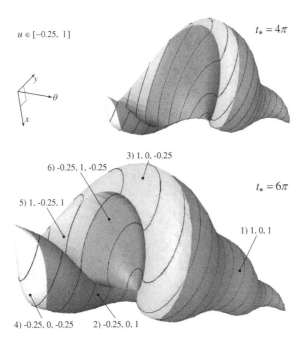

Fig. 24. The reachable sets $G^3(t_*)$ for the instants $t_* = 4\pi$ and $t_* = 6\pi$ for non-symmetric constraint on the control u

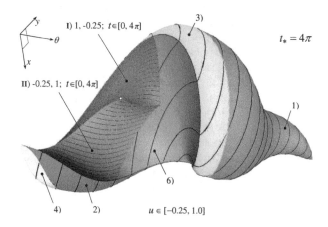

Fig. 25. The reachable set $\mathcal{G}^3(4\pi)$ for non-symmetric constraint on the control u

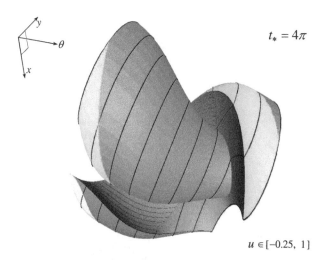

Fig. 26. The reachable set $\mathcal{G}^3(4\pi)$ with the angle θ computed modulo 2π

5. Conclusion

The paper considers reachable sets and inherent character of their development for simplest models of "car" motion used in the mathematical literature. Our investigation is restricted to the cases where reachable sets are constructed in two- and three-dimensional spaces. The understanding of the features and complexities that appear in low dimensional problems can be useful for the analysis of more complex models and for solving real practical problems (Laumond, 1998),(Lensky & Formal'sky, 2003), (LaValle, 2006), (Martynenko, 2007).

6. Acknowledgement

This work is partially supported by the Russian Foundation for Basic Research (project nos. 09-01-00436 and 10-01-96006) and by the Program "Mathematical control theory" of the Presidium of RAS (Ural Branch project 09-П-1-1015).

7. References

Agrachev, A. A. & Sachkov, Yu. (2004). *Control Theory from the Geometric Viewpoint*, Springer, Berlin.

Bakolas, E. & Tsiotras, P. (2011). Optimal synthesis of the asymmetric sinistral/dextral Markov–Dubins problem, *Journal of Optimization Theory and Applications*, Vol. 150 (No. 2), 233–250.

Bardi, M. & Capuzzo-Dolcetta, I. (1997). *Optimal Control and Viscosity Solutions of Hamilton–Jacobi–Bellman Equations. Systems & Control: Foundations and Applications*. Birkhäuser, Boston.

Boissonnat, J.-D. & Bui, X.-N. (1994). *Accessibility region for a car that only moves forwards along optimal paths*, Rapport de recherche N° 2181, INRIA.

Cockayne, E. J. & Hall, G. W. C. (1975). Plane motion of a particle subject to curvature constraints, *SIAM Journal on Control and Optimization*, Vol. 13 (No. 1), 197–220.

Cristiani, E. & Falcone, M. (2009). Fully-discrete schemes for the value function of pursuit-evasion games with state constraints, *Advances in Dynamic Games and Their Applications, Ann. Internat. Soc. Dynam. Games*, Vol. 10, Birkhäuser, Boston, MA, 177–206.

Dubins, L. E. (1957). On curves of minimal length with a constraint on average curvature and with prescribed initial and terminal positions and tangents, *American Journal of Mathematics*, Vol. 79, 497–516.

Grigor'eva, S. V.; Pakhotinskikh, V. Yu.; Uspenskii, A. A.; Ushakov, V. N. (2005). Construction of solutions in some differential games with phase constraints, *Sbornik: Mathematics*, Vol. 196 (No. 3-4), 513–539.

Isaacs, R. (1951). *Games of pursuit*, Scientific report of the RAND Corporation, Santa Monica.

Isaacs, R. (1965). *Differential Games*, John Wiley, New York.

Laumond, J.-P. (ed.) (1998). *Robot Motion Planning and Control*, Lecture Notes in Control and Information Sciences, Vol. 229, Springer, New York.

LaValle, S. M. (2006). *Planning Algorithms*, Chapters 13, 15, Cambridge University Press.

Lee, E. B. & Markus, L. (1967). *Foundations of Optimal Control Theory*, Wiley, New York.

Lensky, A. V. & Formal'sky, A. M. (2003). Two-wheel robot-bicycle with a gyroscopic stabilizer, *Journal of Computer and Systems Sciences International*, No. 3, 482–489.

Markov, A. A. (1889). Some examples of the solution of a special kind of problem on greatest and least quantities, *Soobscenija Charkovskogo Matematiceskogo Obscestva*, Vol. 2-1 (No. 5,6), 250–276 (in Russian).

Martynenko, Yu. G. (2007). Motion control of mobile wheeled robots, *Journal of Mathematical Sciences (New York)*, Vol. 147 (No. 2), 6569–6606.

Mikhalev, D. K. & Ushakov, V. N. (2007). On two algorithms for the approximate construction of a positional absorption set in a game-theoretic approach problem, *Automation & Remote Control*, Vol. 68 (No. 11), 2056–2070.

Patsko, V. S.; Pyatko, S. G.; Fedotov, A. A. (2003). Three-dimensional reachability set for a nonlinear control system, *Journal of Computer and Systems Sciences International*, Vol. 42 (No. 3), 320–328.

Patsko, V. S. & Turova, V. L. (2009). From Dubins' car to Reeds and Shepp's mobile robot, *Computing and Visualization in Science*, Vol. 12 (No. 7), 345–364.

Pecsvaradi, T. (1972). Optimal horizontal guidance law for aircraft in the terminal area, *IEEE Transactions on Automatic Control* Vol. 17 (No. 6), 763–772.

Pontryagin, L. S.; Boltyanskii, V. G.; Gamkrelidze, R. V.; Mischenko E. F. (1962). *The Mathematical Theory of Optimal Processes*, Interscience, New York.

Reeds, J. A. & Shepp, L. A. (1990). Optimal paths for a car that goes both forwards and backwards, *Pacific Journal of Mathematics*, Vol. 145, 367–393.

Sethian, J. A. (1999) *Level Set Methods and Fast Marching Methods*, Cambridge University Press, Cambridge, UK.

Souères, P.; Fourquet, J.-Y.; Laumond, J.-P. (1994). Set of reachable positions for a car, *IEEE Transactions on Automatic Control*, Vol. 39 (No. 8), 1626–1630.

Souères, P. & Laumond, J. P. (1996). Shortest paths synthesis for a car-like robot, *IEEE Transactions on Automatic Control*, Vol. 41 (No. 5), 672–688.

Sussmann, H.J. & Tang, W. (1991). *Shortest paths for the Reeds-Shepp car: a worked out example of the use of geometric techniques in nonlinear optimal control*, Report SYCON-91-10, Rutgers University.

Takei, R. & Tsai, R. (2010). *Optimal trajectories of curvature constrained motion in the Hamilton-Jacobi formulation*, Report 10-67, University of California, Los Angeles.

10

Motion Planning for Mobile Robots Via Sampling-Based Model Predictive Optimization

Damion D. Dunlap[1], Charmane V. Caldwell[2], Emmanuel G. Collins[2],
Jr. and Oscar Chuy[2]
[1]*Naval Surface Warfare Center - Panama City Division*
[2]*Center for Intelligent Systems, Control and Robotics (CISCOR)*
FAMU-FSU College of Engineering
U.S.A

1. Introduction

Path planning is a method that determines a path, consecutive states, between a start state and goal state, LaValle (2006). However, in motion planning that path must be parameterized by time to create a trajectory. Consequently, not only is the path determined, but the time the vehicle moves along the path. To be successful at motion planning, a vehicle model must be incorporated into the trajectory computation. The motivation in utilizing a vehicle model is to provide the opportunity to predict the vehicle's motion resulting from a variety of system inputs. The kinematic model enforces the vehicle kinematic constraints (i.e. turn radius, etc.), on the vehicle that limit the output space (state space). However, the kinematic model is limited because it does not take into account the forces acting on the vehicle. The dynamic model incorporates more useful information about the vehicle's motion than the kinematic model. It describes the feasible control inputs, velocities, acceleration and vehicle/terrain interaction phenomena. Motion planning that will require the vehicle to perform close to its limits (i.e. extreme terrains, frequent acceleration, etc.) will need the dynamic model. Examples of missions that would benefit from using a dynamic model in the planning are time optimal motion planning, energy efficient motion planning and planning in the presence of faults, Yu et al. (2010).

Sampling-based methods represent a type of model based motion planning algorithm. These methods incorporate the system model. There are current sampling-based planners that should be discussed: The Rapidly-Exploring Random Tree (RRT) Planner, Randomized $A^\star$ ($RA^\star$) algorithm, and the Synergistic Combination of Layers of Planning (SyCLoP) multi-layered planning framework. The Rapidly-Exploring Random Tree Planner was one of the first single-query sampling based planners and serves as a foundation upon which many current algorithms are developed. The RRT Planner is very efficient and has been used in many applications including manipulator path planning, Kuffner & LaValle. (2000), and robot trajectory planning, LaValle & Kuffner (2001). However, the RRT Planner has the major drawback of lacking any sort of optimization other than a bias towards exploring the search space. The $RA^\star$ algorithm, which was designed based on the RRT Planner, addresses this drawback by combining the RRT Planner with an $A^\star$ algorithm. The SyCLoP framework is presented because it not only represents a very current sampling-based planning approach, but the framework is also one of the few algorithms to directly sample the control inputs.

Originally, this research began by applying nonlinear model predictive control (NMPC), implemented with sequential programming, to generate a path for an autonomous underwater vehicle (AUV), Caldwell et al. (2006). As depicted in Fig. 1, NMPC was attractive because it is an online optimal control method that incorporates the system model, optimizes a cost function and includes current and future constraints all in the design process. These benefits made planning with NMPC promising, but there were weaknesses of NMPC that had to be addressed. Since MPC must solve the optimization problem online in real-time, the method was limited to slow systems. Additionally, even though models were used in the design process, linear models where typically used in order to avoid the local minima problem that accompany the use of nonlinear models. In order to exploit the benefits of MPC these issues had to be addressed.

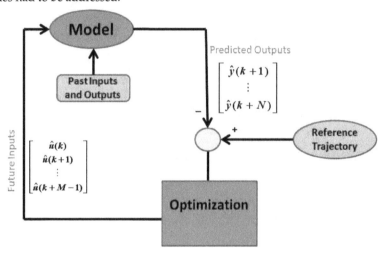

Fig. 1. The stages of the MPC algorithm.

Since the robotics and AI communities had the same goal for planning but have different approaches that tend to yield computationally efficient algorithms, it was decided to integrate these various concepts to produce a new enhanced planner called Sampling Based Model Predictive Control (SBMPC). The concept behind SBMPC was first presented in Dunlap et al. (2008). Instead of utilizing traditional numerical methods in the NMPC optimization phase in Fig. 1, Sampling Based Model Predictive Optimization (SBMPO) uses A^* type optimization from the AI community. This type of graph search algorithm results in paths that do not become stuck in local minima. In addition, the idea of using sampling to consider only a finite number of solutions comes from robotic motion planning community. Sampling is the mechanism used to trade performance for computational efficiency. Instead of sampling in the output space as traditional sampling based planning methods, SBMPC follows the view of traditional MPC and SyCLoP, which samples the input space. Thus, SBMPC draws from the control theory, robotics and AI communities.

Section 2 of this chapter will present the novel SBMPC algorithm in detail and compare Sampling Based Model Predictive Optimization and traditional Sampling based methods. Section 3 provides simulation results utilized on an AUV kinematic model. Section 4 presents results of both an AUV and an unmanned ground vehicle (UGV) that perform steep hill climbing. An evaluation of SBMPO tuning parameters on the computation time and cost is presented in Section 5. Finally, Section 6 concludes and presents future work.

2. SBMPC algorithm

This section provides the SBMPC Algorithm and the comparison of SBMPO to other traditional sampling based methods. However, first the variables used in the algorithm are described. The SBMPO algorithm and terms follow closely with Lifelong Planning A^* (LPA^*), Koenig et al. (2004). However, the variation is in the Generate Neighbor algorithm which generates the next state by integrating the system model and considering constraint violations. All the components of SBMPC are described in this Section, but the later simulation results in Section 3 and 4 utilize only the SBMPO and Generate Neighbors algorithms.

2.1 SBMPC variables

SBMPC operates on a dynamic directed graph G which is a set of all nodes and edges currently in the graph. $SUCC(n)$ represents the set of successors (children) of node $n \in G$ while $PRED(n)$ denotes the set of all predecessors (parents) of node $v \in G$. The cost of traversing from node n' to node $n \in SUCC(n')$ is denoted by $c(n',n)$, where $0 < c(n',n) < \infty$. The optimization component is called Sampling Based Model Predictive Optimization and is an algorithm that determines the optimal cost (i.e. shortest path, shortest time, least energy, etc.) from a start node $n_{start} \in G$ to a goal node $n_{goal} \in G$.

The start distance of node $v \in G$ is given by $g^*(v)$ which is the cost of the optimal path from the given start node v_{start} to the current node v. SBMPC maintains two estimates of $g^*(v)$. The first estimate $g(v)$ is essentially the current cost from v_{start} to the node v while the second estimate, $rhs(v)$, is a one-step lookahead estimate based on $g(v')$ for $v' \in PRED(v)$ and provides more information than the estimate $g(v)$. The $rhs(v)$ value satisfies

$$rhs(v) = \begin{cases} 0, & \text{if } v = v_{start} \\ \min_{v' \in PRED(v)}(g(v') + c(v',v)), & \text{otherwise.} \end{cases} \tag{1}$$

A node v is locally consistent iff $g(v) = rhs(v)$ and locally inconsistent iff $g(v) \neq rhs(v)$. If all nodes are locally consistent, then $g(v)$ satisfies (1) for all $v \in G$ and is therefore equal to the start distance. This enables the ability to trace the shortest path from v_{start} to any node v by starting at v and traversing to any predecessor v' that minimizes $g(v') + c(v',v)$ until v_{start} is reached.

To facilitate fast re-planning, SBMPC does not make every node locally consistent after an edge cost change and instead uses a heuristic function $h(v,v_{goal})$ to focus the search so that it only updates $g(v)$ for nodes necessary to obtain the optimal cost. The heuristic is used to approximate the goal distances and must follow the triangle inequality: $h(v_{goal},v_{goal}) = 0$ and $h(v,v_{goal}) \leq c(v,v') + h(v',v_{goal})$ for all nodes $v \in G$ and $v' \in SUCC(s)$. SBMPO employs the heuristic function along with the start distance estimates to rank the priority queue containing the locally inconsistent nodes and thus all the nodes that need to be updated in order for them to be locally consistent. The priority of a node is determined by a two component key vector:

$$key(v) = \begin{pmatrix} k_1(v) \\ k_2(v) \end{pmatrix} = \begin{pmatrix} \min(g(v),rhs(v)) + h(v,v_{goal}) \\ \min(g(v),rhs(v)) \end{pmatrix} \tag{2}$$

where the keys are ordered lexicographically with the smaller key values having a higher priority.

2.2 SBMPC algorithm

The SBMPC algorithm is comprised of three primary methods: Sampling Based Model Predictive Control, Sampling Based Model Predictive Optimization and Generate Neighbor. The main SBMPC algorithm follows the general structure of MPC where SBMPC repeatedly computes the optimal path between the current state $x_{current}$ and the goal state x_{goal}. After a single path is generated, $x_{current}$ is updated to reflect the implementation of the first control input and the graph G is updated to reflect any system changes. These steps are repeated until the goal state is reached.

The second algorithm SBMPO is the optimization phase of SBMPC that provides the prediction paths. SBMPO repeatedly generates the neighbors of locally inconsistent nodes until v_{goal} is locally consistent or the key of the next node in the priority que is not smaller than $key(v_{goal})$. This follows closely with the ComputeShortestPath algorithm of $LPA^{\star}$ Koenig et al. (2004). The node, v_{best}, with the highest priority (lowest key value) is on top of the priority que. The algorithm then deals with two potential cases based on the consistency of the expanded node v_{best}. If the node is locally overconsistent, $g(v) > rhs(v)$, the g-value is set to $rhs(v)$ making the node locally consistent. The successors of v are then updated. The update node process includes recalculating $rhs(v)$ and key values, checking for local consistency and either adding or removing the node from the priority que accordingly. For the case when the node is locally underconsistent, $g(v) < rhs(v)$, the g-value is set to ∞ making the node either locally consistent or overconsistent. This change can affect the node along with its successors which then go through the node update process.

The Generate Neighbor algorithm determines the successor nodes of the current node. In the input space, a set of quasi-random samples are generated that are then used with a model of the system to predict a set of paths to a new set of outputs (nodes) with $x_{current}$ being the initial condition. The branching factor B (sampling number) determines the number of paths that will be generated and new successor nodes. The path is represented by a sequence of states $x(t)$ for $t = t_1, t_1 + \Delta t, \cdots, t_2$, where Δt is the model step size. The set of states that do not violate any state or obstacle constraints is called $\mathbf{X}_{free}$. If $x(t) \in \mathbf{X}_{free}$, then the new neighbor node x_{new} and the connecting edge can be added to the directed graph, G. If $x_{new} \in STATE\_GRID$, then the node currently exists in the graph and only the new path to get to the existing node needs to be added.

Algorithm 1 Sampling Based Model Predictive Control

1: $x_{current} \Leftarrow$ **start**
2: **repeat**
3:      SBMPO ( )
4:      Update system state, $x_{current}$
5:      Update graph, **G**
6: **until** the goal state is achieved

2.3 Comparison of SBMPO and traditional sampling based methods

This section discusses the conceptual comparison between SBMPO and traditional Sampling-based methods. Similar to many other planning methods, there have been many variants of the sampling based methods that seek to improve various aspects of their

---

Algorithm 2 SBMPO ( )

1: **while** $PRIORITY.TopKey() < v_{goal}.key \parallel v_{goal}.rhs \neq v_{goal}.g$ **do**
2:     $v_{best} \Leftarrow PRIORITY.Top()$
3:     Generate_Neighbors (v_{best}, **B**)
4:     **if** $v_{best}.g > v_{best}.rhs$ **then**
5:        $v_{best}.g = v_{best}.rhs$
6:        **for all** $v \in SUCC_{v_{best}}$ **do**
7:           Update the node, v
8:        **end for**
9:     **else**
10:        $v_{best}.g = \infty$
11:        **for all** $v \in SUCC(v_{best}) \cup v_{best}$ **do**
12:           Update the node, v
13:        **end for**
14:     **end if**
15: **end while**

Algorithm 3 Generate_Neighbors (Vertex v, Branching **B**)

1: **for** $i = 0$ to **B** **do**
2:     Generate sampled input, $u \in \mathbb{R}^u \cap \mathbf{U}_{free}$
3:     **for** $t = t_1 : dt_{integ} : t_2$ **do**
4:        Evaluate model: $x(t) = f(v.x, u)$
5:        **if** $x(t) \notin \mathbf{X}_{free}(t)$ **then**
6:           **Break**
7:        **end if**
8:     **end for**
9:     $x_{new} = x(t_2)$
10:     **if** $x_{new} \in STATE\_GRID$ and $x_{new} \in \mathbf{X}_{free}$ **then**
11:        Add $Edge(v.x, x_{new})$ to graph, **G**
12:     **else if** $x_{new} \in \mathbf{X}_{free}$ **then**
13:        Add $Vertex(x_{new})$ to graph, **G**
14:        Add $Edge(v.x, x_{new})$ to graph, **G**
15:     **end if**
16: **end for**

---

performance. It is not possible to cover every variant, but the purpose of this section is to put in perspective how SBMPO is a variant of the traditional sampling-based method.

2.3.1 Traditional sampling based methods

Examples of traditional sampling based motion planning algorithms include RRTs, LaValle (1998), and probability roadmaps,. A common feature of each of these algorithms is they work in the output space of the robot and employ various strategies for generating samples (i.e., random or pseudo-random points). In essence, as shown in Fig. 2, sampling based motion planning methods work by using sampling to construct a tree that connects the root (initial state) with a goal region.

Most online sampling based planning algorithms follow this general framework:

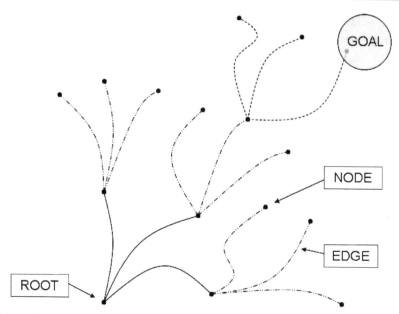

Fig. 2. A tree that connects the root with a goal region.

1. **Initialize:** Let $G(V; E)$ represent a search graph where V contains at least one vertex (i.e., node), typically the start vertex and E does not contain any edges.

2. **Vertex Selection Method (VSM):** Select a vertex u in V for expansion.

3. **Local Planning Method (LPM):** For some $u_{new} \in C_{free}$ (free states in the configuration space) and attempt to generate a path $\tau_s : [0,1] \rightarrow: \tau(0) = u$ and $\tau(1) = u_{new}$. The path must be checked to ensure that no constraints are violated. If the LPM fails, then go back to Step 2.

4. **Insert an Edge in the Graph:** Insert τ_s into E, as an edge from u to u_{new}. Insert u_{new} into V if it does not already exist.

5. **Check for a Solution:** Check G for a solution path.

6. **Return to Step 2:** Repeat unless a solution has been found or a failure condition has been met.

The model is part of the local planning method (LPM), which determines the connection between the newly generated sample and the existing graph. Essentially, it is a two-point value boundary problem.

2.3.2 Similarities of SBMPO and traditional sampling based methods
There are some similarities that both SBMPO and traditional sampling methods share.

2.3.2.1 Sampling

As its name implies, SBMPC is dependent upon the concept of sampling, which has arisen as one of the major paradigms for robotic motion planning community, LaValle (2006). Sampling is the mechanism used to trade performance for computational efficiency. SBMPO employs quasi-random samples of the input space. Properly designed sampling algorithms provide theoretical assurances that if the sampling is dense enough, the sampling algorithm will find a solution when it exists (i.e. it has some type of completeness).

2.3.3 Differences of SBMPO and traditional sampling based methods

Since SBMPO is the outgrowth of both MPC and graph search algorithms, there are some fundamental differences in SBMPO and traditional sampling based methods.

2.3.3.1 Input sampling

There are two primary disadvantages to using output (i.e., configuration space) sampling as is commonly done in traditional sampling based methods. The first limitation lies within the VSM, where the algorithm must determine the most ideal node to expand. This selection is typically made based on the proximity of nodes in the graph to a sampled output node and involves a potentially costly nearest neighbor search. The LPM presents the second and perhaps more troublesome problem, which is determining an input that connects a newly sampled node to the current node. This problem is essentially a two-point boundary value problem (BVP) that connects one output or state to another. There is no guarantee that such an input exists. Also, for systems with complex dynamics, the search itself can be computationally expensive, which leads to a computationally inefficient planner. A solution to the problem is to introduce input sampling. The concept of input sampling is not new and has been integrated into methods like the SyCLoP algorithm, Plaku et al. (2010). When the input space is sampled as proposed in this chapter, the need for a nearest-neighbor search is eliminated, and the LPM is reduced to the integration of a system model, and therefore, only generates outputs that are achievable by the system. Sampling the control inputs directly also prevents the need to determine where to connect new samples to the current graph and therefore avoid costly nearest-neighbor searches.

In order to visualize this concept, consider an Ackerman steered vehicle at rest that has position (x, y) and orientation θ, which are the outputs of the kinematic model. The model restricts the attainable outputs. All the dots in Fig. 3 are output nodes obtained from sampling the output space even though only the dots on the mesh surface can physically be obtained by the vehicle. There are a larger number of dots (sampled outputs) in the output space that do not lie in the achievable region (mesh surface). This means those sampled outputs are not physically possible, so traditional sample based methods would have to start the search over. This leads to an inefficient search that can substantially increase the computational time of the planner. The intersection of the grid lines in Fig. 3 correspond to the points in output space generated by a uniform sampling of the model inputs, the left and right wheel velocities. In essence, sampling in the input space leads to more efficient results since each of the corresponding dots in the output space is allowed by the model.

2.3.3.2 Implicit state grid

Although input sampling avoids two of the primary computational bottle-necks of sampling-based motion planning, there is also a downside of input sampling. Input sampling has not been used in most planning research, because it is seen as being inefficient. This type of sampling can result in highly dense samples in the output space since input sampling does not inherently lead to a uniformly discretized output space, such as a uniform grid. This problem is especially evident when encountering a local minimum problem associated with the $A^\star$ algorithm, which can occur when planning in the presence of a large concave obstacle while the goal is on the other side of the obstacle. This situation is considered in depth for discretized 2D path planning in the work of Likhachev & Stentz (2008), which discusses that

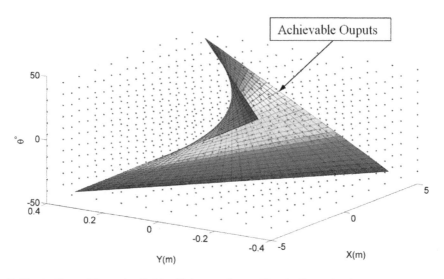

Fig. 3. Illustration of the potential inefficiency of sampling in the output space.

the $A^\star$ algorithm must explore all the states in the neighborhood of the local minimum, shown as the shaded region of Fig. 4, before progressing to the final solution. The issue that this presents to input sampling methods is that the number of states within the local minimum is infinite because of the lack of a discretized output space.

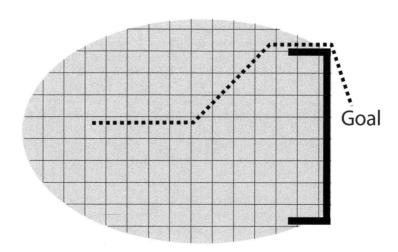

Fig. 4. Illustration of the necessity of an implicit state grid.

The second challenge resulting from the nature of input sampling as well as the lack of a grid is that the likelihood of two outputs (states) being identical is extremely small. All $A^\star$-like algorithms utilize Bellman's optimality principle to improve the path to a particular output by updating the paths through that output when a lower cost alternative is found. This feature is essential to the proper functioning of the algorithm and requires a mechanism to identify when outputs (states) are close enough to be considered the same. The scenario presented in Fig. 5 is a situation for which the lack of this mechanism would generate an inefficient path. In this situation, node v_1 is selected for expansion after which the lowest cost node is v_3. The *implicit state grid* then recognizes that v_2 and v_3 are close enough to be considered the same and updates the path to their grid cell to be path c since $c < a + b$.

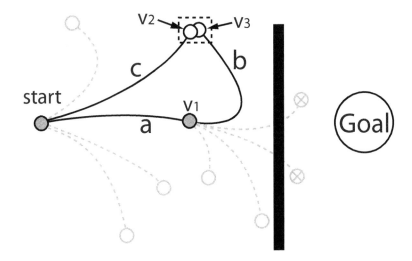

Fig. 5. Illustration of the necessity of an implicit state grid.

The concept of an *implicit state grid*, Ericson (2005), is introduced as a solution to both of the challenges generated by input sampling. The implicit grid ensures that the graph generated by the SBMPC algorithm is constructed such that only one active output (state) exists in each grid cell, limiting the number of nodes that can exist within any finite region of the output space. In essence, the *implicit state grid* provides a discretized output space. It also allows for the efficient storage of potentially infinite grids by only storing the grid cells that contain nodes, which is increasingly important for higher dimensional problems, Ericson (2005). The resolution of the grid is a significant factor in determining the performance of the algorithm with more fine grids in general requiring more computation time, due to the increased number of outputs, with the benefit being a more optimal solution. Therefore, the grid resolution is a useful tuning tool that enables SBMPC to effectively make the trade off between solution quality and computational performance.

2.3.3.3 Goal directed optimization

There is a class of discrete optimization techniques that have their origin in graph theory and have been further developed in the path planning literature. In this study these techniques will be called *goal-directed optimization* and refer to graph search algorithms such as Dijkstra's algorithm and the $A^\star$, $D^\star$, and $LPA^\star$ algorithms Koenig et al. (2004); LaValle (2006). Given a graph, these algorithms find a path that optimizes some cost of moving from a start node to some given goal. In contrast to discrete optimization algorithms such as branch-and-bound optimization Nocedal & Wright (1999), which "relaxes" continuous optimization problems, the goal-directed optimization methods are inherently discrete, and have often been used for real-time path planning.

Generally, sampling based methods such as RRTs do not incorporate any optimization and terminate when an initial feasible solution is determined. In essence, instead of determining an optimal trajectory, traditional sampling based methods only attempt to find feasible trajectories. To remedy these problems, the Randomized $A^\star$ ($RA^\star$) algorithm was introduced in Diankov & Kuffner. (2007), as a hybrid between RRTs and the $A^\star$ search algorithm. Similar to $RA^\star$, SBMPO incorporates a goal directed optimization to ensure the trajectory is optimal subject to the sampling.

Although not commonly recognized, goal-directed optimization approaches are capable of solving control theory problems for which the ultimate objective is to plan an optimal trajectory and control inputs to reach a goal (or set point) while optimizing a cost function. Hence, graph search algorithms can be applied to terminal constraint optimization problems and set point control problems. To observe this, consider the tree graph of Fig. 2. Each node of this tree can correspond to a system state, and the entire tree may be generated by integrating sampled inputs to a system model. Assume that the cost of a trajectory is given by the sum of the cost of the corresponding edges (i.e., branches), where the cost of each edge is dependent not only on the states it connects but also the inputs that are used to connect those states. The use of the system model can be viewed simply as a means to generate the directed graph and associated edge costs.

3. 3D motion planning with kinematic model

In order to demonstrate SBMPO capabilities, two local minima scenarios will be considered: 1) a concave obstacle and 2) a highly cluttered area. The purpose is to test how SBMPO handles these types of local minima environments. In this section, the kinematic model of an AUV is used for the motion planning simulations.

$$
\begin{bmatrix} \dot{x} \\ \dot{y} \\ \dot{z} \end{bmatrix} = \begin{bmatrix} c\theta c\psi & s\phi s\theta c\psi - c\phi s\psi & c\phi s\theta c\psi - s\phi s\psi \\ c\theta s\psi & s\phi s\theta s\psi - c\phi c\psi & c\phi s\theta s\psi - s\phi c\psi \\ s\theta & s\phi c\theta & c\phi c\theta \end{bmatrix} \begin{bmatrix} u \\ v \\ w \end{bmatrix}
$$

$$
\begin{bmatrix} \dot{\phi} \\ \dot{\theta} \\ \dot{\psi} \end{bmatrix} = \begin{bmatrix} 1 & s\phi t\theta & c\phi t\theta \\ 0 & c\phi & -s\phi \\ 0 & s\phi s\theta & c\phi s\theta \end{bmatrix} \begin{bmatrix} p \\ q \\ r \end{bmatrix}, \tag{3}
$$

where u, v, w are linear velocities in the local body fixed frame along the x, y, z axes, respectively and p, q, r are the angular velocities in the local body fixed frame along the x, y, z axes, respectively. The AUV posture can be defined by six coordinates, three representing the position $x_1 = (x, y, z)^T$ and three corresponding to the orientation $x_2 = (\phi, \theta, \psi)^T$, all with respect to the world frame. The constraints for the vehicle is given in Table 1.

The basic problem in each of these scenarios is to use the kinematic model to plan a minimum distance trajectory for the AUV from a start posture to a goal position while avoiding the obstacles. A 2.93 GHz Intel Core 2 Duo desktop was used for simulations in this Section.

3.1 AUV concave obstacle

As previously stated, SBMPO can handle local minimum problems that other path planning methods have difficulties handling. A local minima problem is a possible scenario a vehicle

Inputs	min	max	States	min	max
u	0 m/s	2 m/s	x	-5 m	30 m
v	-0.1 m/s	0.1 m/s	y	-5 m	30 m
w	-0.1 m/s	0.1 m/s	z	-20 m	0 m
p	$-5°/s$	$5°/s$	ϕ	$-15°$	$15°$
q	$-5°/s$	$5°/s$	θ	$-15°$	$15°$
r	$-15°/s$	$15°/s$	ψ	$-360°$	$360°$

Table 1. Simulation Constraints for the 3D Kinematic Model

can be presented with that has a set of concave obstacles in front of the goal. Note that whenever a vehicle is behind an obstacle or group of obstacles and has to increase its distance from the goal to achieve the goal, it is in a local minimum position.

The simulations were run with a sampling number of 25 and grid resolution of 0.1m. The vehicle has a start posture of $(5m, 0m, -10m, 0°)$ and a goal position of $(5m, 10m, -10m)$. As shown in Fig. 6, SBMPO does not get stuck behind the obstacles, but successfully determines a trajectory in 0.59s. The successful traversal is largely due to the optimization method used in SBMPO. The goal-directed optimization allows a more promising node (lower cost) to replace a higher cost node as shown in the the example in Fig 7. Goal-directed optimization can accomplish this because they compute each predicted control input separately and backs up when needed as illustrated by the iterations corresponding to the 3rd, 4th, and 5th arrays. This feature enables it to avoid local minima. It converges to the optimal predicted control sequence $\{u^*(k)\}$ (denoted by the right-most array), which is the optimal solution subject to the sampling, whereas a nonlinear programming method may get stuck at a local minimum. In addition, these results show that SBMPO's implementation of the implicit state grid helps prevent the issues with input sampling discussed in Section 2.3.3.2. Since the implicit grid is applied, it does not require significant time to explore the area around the concave obstacles.

3.2 AUV cluttered multiple obstacles

In Section 3.1, there was one local minimum in the scenario. However, some underwater environments will require the AUV to navigate around multiple cluttered obstacles. This can produce a more complex situation because now there are multiple local minima . The simulations in this section assume there were random start, goal and obstacle locations in order to represent 100 different multiple obstacle underwater environment configurations. The start locations X, Y, Z and ψ were chosen randomly in the respective ranges $[0 \ 20]m$, $[0 \ 1]m$, $[-12 \ -8]m$, $[30° \ 150°]$, and the goal was chosen randomly in the respective ranges $[0 \ 20]m$, $[19 \ 20]m$, $[-12 \ -8]m$. In addition, there were 40 randomly generated obstacles. The 100 simulation runs had a sampling number of 25. Fig. 8 exemplifies one random scenarios generated. In the scenarios SBMPO was capable of allowing the AUV to maneuver in the cluttered environment successfully reaching the goal.

For a vehicle to be truly autonomous, it must be capable of determining a trajectory that will allow it to successfully reach the goal without colliding with an obstacle. In these simulations SBMPO was 100% successful in assuring the vehicle accomplished this task. It is important to consider both SBMPO's mean computation time of 0.43s and median computation time of 0.15s to compute these trajectories. Since the scenarios generated were random, there were a few scenarios created that were more cluttered which caused a larger CPU time. This is the reason for the discrepancy between the mean and median computation times.

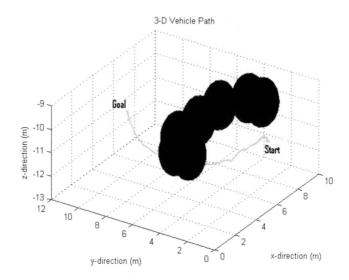

Fig. 6. A local minima scenario.

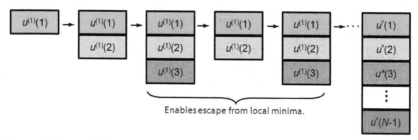

Fig. 7. Example of goal-directed optimization.

4. 3D motion planning with dynamic model

In some scenarios it is sufficient to plan using the kinematic model. However, in cases where the vehicle is pushed to an extreme, it is necessary to consider the vehicles dynamic model when planning.

4.1 Steep hill climbing

In this section, two different types of vehicles, an AUV and an UGV, consider steep hill motion planning using their respective dynamic model. The vehicle must be capable of acquiring a certain amount of momentum to successfully traverse the hill. In order to determine if the vehicle can produce the correct amount of momentum, a dynamic model is not physically capable of traversing.

4.1.1 AUV

The AUV dynamic model used for these simulations can be found in Healey & Lienard (1993). The model constraints are given in Table 2. The SBMPO parameters are in Table 3. The steep

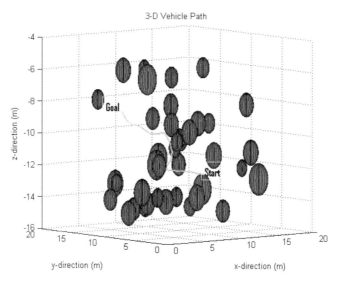

Fig. 8. A random start, goal and obstacle scenario.

hill was constructed utilizing 6 sphere obstacles constraints stacked to give a peak at $8m$. The AUV's start location was $(-4m\ 17m\ -18m\ 0°\ 0°\ 0°)$ and goal was $(19m\ 7m\ -14m)$.

States	min	max	States	min	max	Inputs	min	max
x	-30 m	130 m	u	0 m/s	2 m/s	δ_r	$-22°$	$22°$
y	-30 m/s	130 m	v	-2 m/s	2 m/s	δ_s	$-22°$	$22°$
z	-20 m	5 m	w	-2 m/s	2 m/s	δ_b	$-22°$	$22°$
ϕ	$-15°$	$15°$	p	$-5°/s$	$5°/s$	δ_{bp}	$-22°$	$22°$
θ	$-85°$	$85°$	q	$-5°/s$	$5°/s$	δ_{bs}	$-22°$	$22°$
ψ	$-360°$	$360°$	r	$-15°/s$	$15°/s$	n	0 rpm	1500 rpm

Table 2. The simulation constraints for the AUV dynamic model.

Model Time Steps	0.5s
Control updates	10s
No. of Input Samples	20
Grid Resolution	0.5

Table 3. The simulation parameters for the AUV dynamic model.

A dynamic model was utilized to determine the path over a steep hill. The AUV was not able to determine a path because the vehicle starts too close to the steep hill to gain momentum. As depicted in Fig. 9, the dynamic model was able to predict that there was not enough momentum to overcome the hill in such a short distance. Note the vehicle constraints do not allow this type of AUV to have a negative velocity which would allow the vehicle to be able to reverse in order to acquire enough momentum. As a result of the vehicle constraint, Fig 9 shows the unsuccessful path. It is not the SBMPO algorithm that cannot successfully determine a path, but the vehicle constraint (dynamic model) that predicts there was not enough momentum to overcome the hill in such a short distance. In order to demonstrate this consider the same scenario using the kinematic model in Fig 10. SBMPO does determine a path, but this is only because the kinematic model utilized does not provide all the vehicle information to correctly predict the vehicle motion. This further shows the importance of using the proper model when motion planning. The trajectory determined by the planner is only as accurate as the model used.

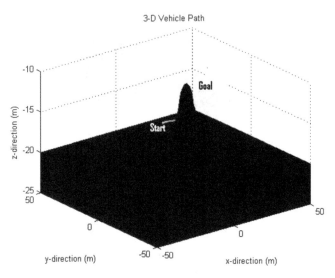

Fig. 9. The AUV dynamic model steep hill scenario.

4.1.2 UGV

This section discusses momentum-based motion planning applied to UGV steep hill climbing. Note that steep hill climbing capability of UGVs is very important to aid the completion of assigned tasks or missions. As an additional requirement, the motion planning is constrained such that the UGV has a zero velocity at the goal (e.g. top of the hill) and this has a unique application, such as reconnaissance, where UGV needs to climb and stop at the top of the hill to gather information. In this section, the momentum-based motion planning is implemented using SBMPO with UGV's dynamic model and a minimum time cost function. The minimum time cost function is employed to achieve zero velocity at the goal.

Figure 11 shows a scenario where a UGV is at the bottom of a steep hill and the task is to climb to the top of the hill. The general approach is to rush to the top of the hill. However, if the torque of the UGV and the momentum are not enough, it is highly possible that the UGV will fail to climb as shown in Fig. 11(a). An alternative approach for the UGV is to back up to gain more momentum and rush to the top of the hill as shown in Fig 11(b). The aforementioned approaches can be done using SBMPO with UGV's dynamic model. SBMPO can generate a trajectory for successful steep hill climbing, and it can also determine if the UGV needs to back up or how far the UGV needs to back up to successfully climb the hill.

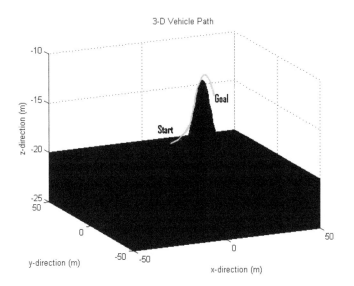

Fig. 10. The AUV kinematic model steep hill scenario.

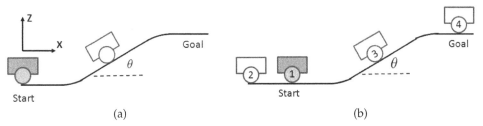

(a)   (b)

Fig. 11. A steep hill climbing scenario for a UGV (a) the UGV rushes to the top of the hill without enough momentum and torque and leads to unsuccessful climb (b) the UGV backs up to gain momentum and leads to a successful climb.

A minimum time cost function is used to implement steep hill climbing with zero velocity at the goal. To formulate the minimum time, consider a system described by

$$\ddot{q} = u; \quad q(0) = q_0, \quad \dot{q}(0) = \omega_0, \tag{4}$$

where u is bounded by $-a \leq u \leq b$. The state space description of (4) is given by

$$\dot{q}_1 = q_2, \quad \dot{q}_2 = u; \quad q_1(0) = q_0 \overset{\Delta}{=} q_{1,0}, \quad q_2(0) = \omega_0 \overset{\Delta}{=} q_{2,0}, \tag{5}$$

where $q_1 = q$ and $q_2 = \dot{q}$. It is desired to find the minimum time needed to transfer the system from the original state $(q_{1,0}, q_{2,0})$ to the final state $(q_{1,f}, 0)$, where $q_{1,f} \overset{\Delta}{=} q_f$. Since the solution for transferring the system from $(q_{1,0}, q_{2,0})$ to the origin $(0, 0)$ is easily extended to the more general case by a simple change of variable, for ease of exposition it is assumed that

$$q_{1,f} = 0. \tag{6}$$

The minimum time control problem described above can be solved by forming the Hamiltonian and applying the "Minimum Principle" (often referred to as "Pontryagin's Maximum Principle") as described in Bryson & Ho (1975). In fact, the above problem is solved in Bryson & Ho (1975) for the case when the parameters a and b are given by $a = b = 1$. Generalizing these results yields that the minimum time is the solution t_f of

$$t_f^2 - \frac{2q_{2,0}}{a}t_f = \frac{q_{2,0}^2 + 2(a+b)q_{1,0}}{ab}, \text{ if } q_{1,0} + \frac{q_{2,0}|q_{2,0}|}{2b} < 0,$$

$$t_f^2 + \frac{2q_{2,0}}{b}t_f = \frac{q_{2,0}^2 - 2(a+b)q_{1,0}}{ab}, \text{ if } q_{1,0} + \frac{q_{2,0}|q_{2,0}|}{2a} > 0. \tag{7}$$

The minimum time (t_f) computed using (7) corresponds to a "bang-bang" optimal controller illustrated by Fig. 12, which shows switching curves that take the system to the origin using either the minimum or maximum control input (i.e., $u = -a$ or $u = b$). Depending on the initial conditions, the system uses either the minimum or maximum control input to take the system to the appropriate switching curve. For example, if $(q_{1,0}, q_{2,0})$ corresponds to point p_1 in Fig. 12, then the control input should be $u = -a$ until the system reaches point p_2 on the switching curve corresponding to $u = b$. At this point the control is switched to $u = b$, which will take the system to the origin.

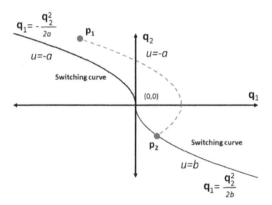

Fig. 12. Illustration of bang-bang minimum time optimal control which yields the minimum time solution t_f of (7).

To demonstrate steep hill climbing, the UGV starts at $(0,0,0)$[m] and the goal is located at $(2.5,0,0.76)$[m]. As shown in Fig. 13, the hill is described with the following parameters: $R = 1m$, $l = 0.75m$, $d = 0.4m$ and $\theta = 30°$. A UGV dynamic model discussed in Yu et al. (2010) is used and it is given by

$$M\ddot{q} + C(\dot{q}, q) + G(q) = \tau, \tag{8}$$

where

$$-\tau_{max} < \tau < \tau_{max} \tag{9}$$

and $\tau_{max} = 10Nm$. $\ddot{q}$, $\dot{q}$, and q are respectively the wheel angular acceleration, velocity, and position, M is the inertia, $C(\dot{q}, q)$ is the friction term, and $G(q)$ is the gravity term. Based on the parameters of the hill and the UGV, the maximum required torque to climb quasi-statically the hill is 14.95Nm. This clearly shows that the UGV cannot climb without using momentum.

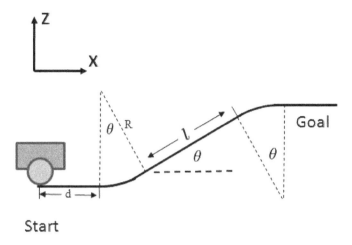

Fig. 13. The UGV steep hill parameters.

The results of the motion planning using SBMPO with the UGV's dynamic model and minimum time cost function are shown in Fig. 14. Fig. 14(a) shows the desired X-Z position of the UGV and Figs. 14(b)-(d) show respectively the desired wheel angular position, velocity, and acceleration, which are the trajectory components of the UGV's. In practice, the resulting trajectory is fed to the UGV's low-level controller for tracking. In Fig. 14(b), the desired wheel angular position starts at zero, and it goes negative (UGV backs up) before it proceeds to the goal. Fig. 14(c) shows the desired angular velocity of the wheel, and it is negative before the UGV accelerates to climb the hill. It also shows that the angular velocity at the goal is zero. The results clearly show that the UGV backs up to increase momentum, which is automatically done by SBMPO.

5. Tuning parameters

Similar to other algorithms, SBMPO has parameters that have to be tuned to guarantee optimal results. SBMPO has two main tuning parameters, the sampling number (branching factor) and grid resolution (size). Each tuning parameter has an effect on the computation time and cost. In this Section, one of the random scenarios from Section 3.2 was investigated.

5.1 Sampling number
The sample number is the number of samples that are selected to span the input space. In order to determine how the sample number effects the computation time of SBMPO the grid resolution was held constant at 0.1m, and the sample number was varied from 10 to 40 by increments of 2. Fig. 15 shows that the effect of the sampling number is nonlinear, so there is no direct relationship between the sample number and computation time. Originally it was thought that an increase in sample number would cause an increase in computation time because there would be more nodes to evaluate. However, as shown in Fig. 15 this is not completely true.

The reason for the nonlinear trend is threefold. First as shown in Fig. 15 by samples 10 and 12 , when there are not enough samples (the sample number is too low) to span the space, it can

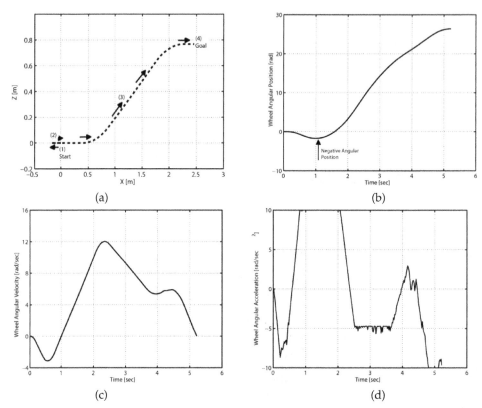

Fig. 14. (a) X-Z position of the UGV (b) wheel angular position (c) wheel angular velocity (d) wheel angular acceleration.

also increase the CPU time, because it takes more iterations (i.e. steps in SBMPO) to determine the solution. A good tuning of the parameter occurs at 14 samples which results in a smaller computation time. The second trend, as shown in Fig. 15 between samples 14 and 22, is that after a good tuning of the parameter, increasing the number of samples also increases the computation times which corresponds to the original hypothesis that an increase in sample number will result in an increase in CPU time. Lastly, a factor that contributes to the effect of the sample number on the computation time is the path produced by SBMPO. It is possible for a larger sample number to have a lower computation time when the path SBMPO generates to the goal encounters a smaller cluster of obstacles. Figs. 16a and 16b show the paths generated respectively using 26 and 36 samples. The path of Fig. 16a which has the lower sampling number takes the AUV through a cluster of obstacles, whereas the path of Fig. 16b which has the larger sample number takes a path that largely avoids the obstacles. Even though Fig. 16b corresponds to a sample number of 36, referring to Fig. 15, its computation time of $0.29s$ is smaller than that for Fig. 16a, which corresponds to a sample number of 26 and has a computation time of $0.5s$.

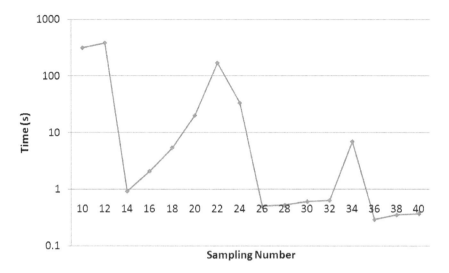

Fig. 15. The effect of sample size on computation time.

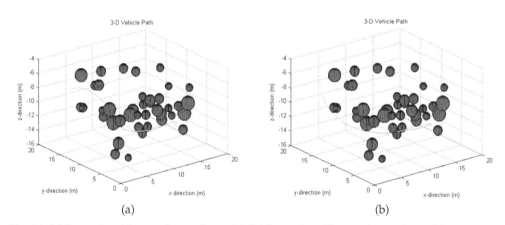

(a)   (b)

Fig. 16. (a) Scenario with sample number = 26, (b) Scenario with sample number = 36.

Fig. 17 depicts how varying the sample number effects the cost (i.e. distance). The cost is larger in the smaller sample numbers 10 and 12. Afterwards, the variation in the cost is small, which leads to more of an optimal solution.

5.2 Grid size
The grid size is the resolution of the implicit state grid. To evaluate how this tuning parameter effects the computation time, the sampling number was held constant at 25, and the grid resolution was varied between 0.02 to 0.5. Again this tuning parameter is not monotonic with respect to the computation time as depicted in Fig. 18. This shows the importance of properly tuning the algorithm. It may be thought that increasing the grid size would cause

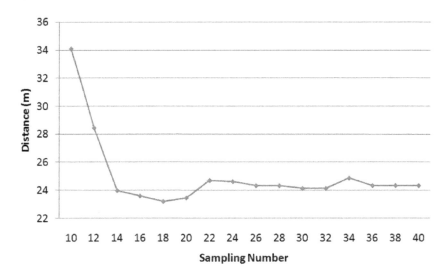

Fig. 17. The effect of sample size on path cost.

less computation. However, the opposite is true. The larger the grid size, the higher the possibility that two nodes are considered as the same state, which leads to the need for more sampling of the input space and an increased computation time. When choosing the grid resolution, it is important to recognize that increasing the grid size tends to lead to higher cost solutions as depicted in Fig. 19.

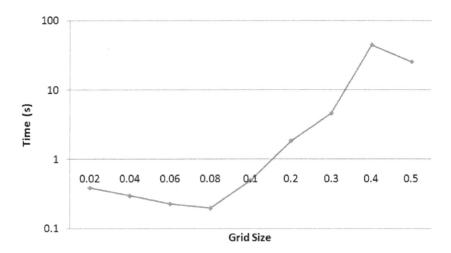

Fig. 18. The effect of grid size on computation time.

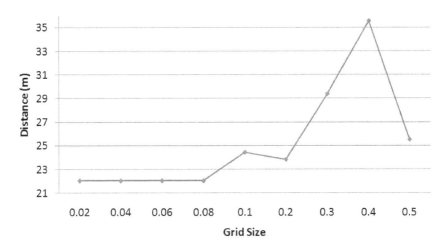

Fig. 19. The effect of grid size on path cost.

6. Conclusion

SBMPO is a NMPC method that exploits sampling-based concepts from the robotics literature along with the $LPA^\star$ incremental optimization algorithm from the AI literature to achieve the goal of quickly and simultaneously determining the control updates and paths while avoiding local minima. The SBMPO solution is globally optimal *subject to the sampling*. Sampling Based Model Predictive Optimization has been shown to effectively generate paths in the presence of nonlinear constraints and when vehicles are pushed to extreme limits. It was determined that SBMPO is only as good as the model supplied to predict the vehicle's movement. Selecting the correct model is important.

The future work is to develop the replanning feature of SBMPC. Currently, SBMPO is applied, but the ability to replan is essential to the SBMPC algorithm. SBMPC utilizes $LPA^\star$ which allows quick replanning of the path without having to completely restart the planning process when new information is obtained or changes in the environment occur. Only the nodes that are affected by a change in the environment must be reevaluated. This reduces the computation time and aids the method in achieving fast computation times. Once the replanning feature of SBMPC is in place, scenarios that include disturbance, model mismatch, unknown obstacles and moving obstacles can be examined to test more realistic situations. The algorithm will also be in a framework that is more comparable to traditional NMPC that only takes the first input and replans at every time step to create a more robust controller. Then SBMPC can be considered a general fast NMPC method that is useful for any nonlinear system or systems subject to nonlinear constraints.

7. References

Bryson, A. & Ho, Y. (1975). *Applied Optimal Control Optimization, Estimation, and Control*, HPC, New York.

Caldwell, C., Collins, E. & Palanki, S. (2006). Integrated guidance and control of AUVs using shrinking horizon model predictive control, *OCEANS Conference* .

Diankov, R. & Kuffner., J. (2007). Randomized statistical path planning, *Conference on Intelligent Robots and Systems* .

Dunlap, D. D., E. G. Collins, J. & Caldwell, C. V. (2008). Sampling based model predictive control with application to autonomous vehicle guidance, *Florida Conference on Recent Advances in Robotics* .

Ericson, C. (2005). *Real–Time Collision Detection*, Elsevier.

Healey, A. & Lienard, D. (1993). Multivariable sliding-mode control for autonomous diving and steering for unmanned underwater vehicle, *IEEE Journal of Oceanic Engineering* 18(3): 327–338.

Koenig, S., Likhachev, M. & Furcy, D. (2004). Lifelong planning A*, *Artificial Intelligence* .

Kuffner, J. J. & LaValle., S. M. (2000). Rrt-connect: An efficient approach to single-query path planning, *IEEE International Conference on Robotics and Automation* p. 995ï£¡1001.

LaValle, S. (1998). Rapidly-exploring random trees: A new tool for path planning, *Technical report*, Iowa State University.

LaValle, S. M. (2006). *Planning Algorithms*, Cambridge University Press.

LaValle, S. M. & Kuffner, J. J. (2001). Randomized kinodynamic planning, *International Journal of Robotics Research* 20(8): 378–400.

Likhachev, M. & Stentz, A. (2008). R* search, *Proceedings of the National Conference on Artificial Intelligence (AAAI)* pp. 1–7.

Nocedal, J. & Wright, S. (1999). *Numerical Optimization*, Springer, New York.

Plaku, E., Kavraki, L. & Vardi, M. (2010). Motion planning with dynamics by synergistic combination of layers of planning, *IEEE Transaction on Robotics* pp. 469–482.

Yu, W., Jr., O. C., Jr., E. C. & Hollis, P. (2010). Analysis and experimental verification for dynamic modeling of a skid-steered wheeled vehicle, *IEEE Transactions on Robotics* pp. 340 – 353.

11

Path Searching Algorithms of Multiple Robot System Applying in Chinese Chess Game

Jr-Hung Guo[1], Kuo-Lan Su[2] and Sheng-Ven Shiau[1]
[1]Graduate school Engineering Science and technology,
National Yunlin University of Science & Technology
[2]Department of Electrical Engineering,
National Yunlin University of Science & Technology
Taiwan

1. Introduction

Chinese chess game [1] is one of the most popular games. A two-player game with a complexity level is similar to Western chess. In the recent, the Chinese chess game has gradually attracted many researcher's attention. The most researchers of the fields are belong to expert knowledge and artificial intelligent. There are many evolutionary algorithms to be proposed. Darwen and Yao proposed the co-evolutionary algorithm to solve problems where an object measure to guide the searching process is extremely difficult to device [2]. Yong proposed multi-agent systems to share the rewards and penalties of successes and failures [3]. Almost all the chess game can be described by game tree. Game tree presents the possible movements and lists all situations for the Chinese chesses. We want to use the multi-robots system to present the scenario of the chess movement for the Chinese chess game, and play the Chinese chess game according to the real-time image feedback to the supervised computer via wireless image system.

The application of co-evolutionary models is to learn Chinese chess strategies, and uses alpha-beta search algorithm, quiescence searching and move ordering [4]. Wang used adaptive genetic algorithm (AGA) to solve the problems of computer Chinese chess [5]. Lee and Liu take such an approach to develop a software framework for rapidly online chess games [6]. Zhou and Zhang present the iterative sort searching techniques based on percentage evaluation and integrate percentage evaluation and iterative sort into problem of Chinese chess computer game [7]. Su and Shiau developed smart mobile robots to speak real-time status using voice module, and program the motion trajectories for multiple mobile robot system [8].

With the robotic technologies development with each passing day, robot system has been widely employed in many applications. Recently, more and more researchers are interest in the robot which can helps people in our daily life, such as entertaining robots, museum docent robots, educational robots, medical robots, service robots, office robots, security robots, home robots, and so on. In the future, we believe that intelligent robots will play an important role in our daily life. In the past literatures, many experts researched in the mobile robot, and proposed many methods to enhance the functions of the mobile robot [9]. So far, developing a big sized mobile robot to be equipped with many functions to become complex and huge, and the development period is too long. Thus, recently small-sized mobile robot system has been investigated for a specific task, and program the optimal motion path on the dynamic environment [18].

There is a growing in multi-robot cooperation research in recent year. Compare to single mobile robot, cooperation multiple mobile robots can lead to faster task completion, higher quality solution, as well as increase robustness owing its ability adjust to robot failure [10]. Grabowski and Navarro-serment [11] suggested multiple mobile robots in which each mobile platform has a specific sensor for some purpose and therefore the system's task can be distributed to each mobile platform during surveillance. The feature of this system is that each mobile robot has a common motion platform, but has different sensors. Chung et al. [12] composed of one ship and four small-sized search robots for team work in hazardous environment. Balch et al. [13] and Alami et al. [14] investigated cooperation algorithm of multiple robot system.

Some papers consider the problem of the multiple robot system working together. The multiple mobile robot system has more advantages than one single robot system [15]. The multiple mobile robots have the potential to finish some tasks faster than a single robot using ant colony system [16]. Multiple mobile robots therefore can be expected more fault tolerant than only one robot. Another advantage of multiple mobile robots is due to merging of overlapping information, which can help compensate for sensor uncertainty [17]. We have developed multiple small-size robot system to be applied in Chinese chess [8]. We extend the application field of the multiple mobile robot system, and program the shortest path moving on the chessboard platform using A* searching algorithm. The A* heuristic function are introduced to improve local searching ability and to estimate the forgotten value [19].

We present the searching algorithms based Chinese chess game, and use multiple mobile robots to present the scenario on the chessboard platform. The mobile robot has the shape of cylinder and its diameter, height and weight is 8cm, 15cm and 1.5kg. The controller of the mobile robot is MCS-51 chip, and acquires the detection signals from sensors through I/O pins, and receives the command from the supervised compute via wireless RF interface. The chesses (mobile robots) can speak Chinese language for real-time status using voice module. We develop the user interface of multiple mobile robots according to the basic rules of Chinese chess game on the supervised computer. The mobile robot receives the command from the supervised computer, and calculates the displacement using the encoder module The supervised computer program the motion trajectories using evaluation algorithm and A* searching algorithm for the mobile robot to play the Chinese chess game. The A* searching algorithm can solve shortest path problem of mobile robots from the start point to the target point on the chessboard platform.

The simulation results can found the shortest motion path for mobile robots (chesses) moving to target points from start points in a collision-free environment. The two points are selected by two players according to the rules of the Chinese chess game. In the experimental results, we test some functions of the mobile robot to obey the moving rules of the game on the chessboard, and implement the simulation results on the chessboard platform using mobile robots. Users can play the Chinese chess game on the supervised computer using the mouse. Mobile robots can receive the command from the supervised computer, and move the next position according to the attribute of the chess. The chess (mobile robot) moving scenario of the Chinese chess game feedback to the user interface using wireless image system.

2. System architecture

The system architecture of the multiple mobile robots based Chinese chess game system is shown in Fig 1. The system contains a supervised computer, a image system, some wireless RF modules, a remote supervised computer and thirty-two mobile robots. Mobile robots are classified two sides (red side and black side) belong two players. There are sixteen mobile robots in each side. The supervised computer can transmit the command signals to control mobile robots, and receives the status of the mobile robots via wireless RF interface. These signals contain ID code, orientation and displacement of mobile robots from the supervised computer to mobile robots. The feedback signals of mobile robots contain ID code, position (X axis and Y axis) to the supervised computer. Each robot is arranged an ID code to classify the attribute of the chess. Users move the chess on the user interface using the mouse. The supervised computer can controls the mobile robot moving the target position according the command. The mobile robot transmits the ending code to the supervised computer after finishing the task via wireless RF interface.

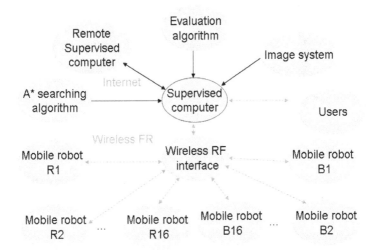

Fig. 1. The architecture of the mobile robots based Chinese chess game system.

The Chinese chess game system can transmit the real-time image to the supervised computer via image system. Users can play the Chinese chess game with others on the supervised computer using the mouse, or play the game on the remote supervised computer via wireless Internet. The player can move the chess piece using the mouse according to the rules of the Chinese chess game. Mobile robots receive the status from the supervised computer via wireless RF interface. There are two sides of the game system. One is the red side; the other is black side. Each side has sixteen chesses. There are one "king" chess, two "advisor" chesses, two "elephant" chesses, two "horse" chesses, two "rook" chesses, two "cannon" chesses and five "pawn" chesses. The definition of the chessboard is shown in Fig. 2. Then we define the initial position all chesses. Such as the position of "red king" is (4,0), and "black king" is (4,9)...etc.

The basic rules of the Chinese chess are found easily on the Internet. Before we want to control the multiple mobile robots based Chinese chess game. A chess engine needs to be designed and tested to ensure that all chess displacement and game rules are strictly adhered. We proposed the engine to be simple programmed basic rules of the games. The chessboard, denoted a 9 by 10 matrix, is the most important information that determine a players next moving position. We use axis position to define all chesses. First, the chessboard, denoted a axis position (x,y) from (0,0) to (8,9).

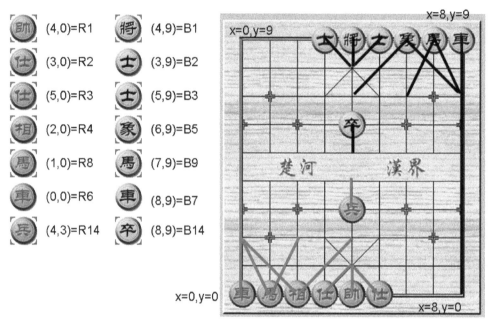

(4,0)=R1   (4,9)=B1

(3,0)=R2   (3,9)=B2

(5,0)=R3   (5,9)=B3

(2,0)=R4   (6,9)=B5

(1,0)=R8   (7,9)=B9

(0,0)=R6   (8,9)=B7

(4,3)=R14   (8,9)=B14

Fig. 2. The positions of the chesses.

We plot the possible motion trajectories using thick lines for the chesses on the chessboard. Then we define the game tree, and move the chess to the target position. Such as the chess "black horse" can moves to the position (8,7), (6,7) or (5,8). But the chess can't moves to the position (5,8) according the rules of the Chinese chess game. The chess "black horse" has an obstacle (black elephant) on the right side, and can't walk over the chessboard. We plot the possible motion trajectory for the chess piece on the chessboard. Then we define the game tree, and move the chess to the assigned position. Such as the chess "red horse" can moves to the position (0,2) or (2,2). But the chess can't moves to the position (3,1) according to the rules of the Chinese chess game.

The communication protocol of the multiple mobile robot system is 10 bytes. It contains one start byte, eight data bytes and one checksum byte. The supervised computer transmits 10 bytes to control the mobile robot, and the communication protocol of the control command is listed in Table 1. The mobile robot receives the command to discriminate the robot ID code to be right, and moves to the target point step by step, and transmits the environment status to the supervised computer on real-time. The communication protocol of the feedback data from the mobile robot is listed in Table 2. The byte 3 and 4 represents the positions of the mobile robot on X axis and Y axis.

Byte	0	1	2	3	4
Definition	Start byte	ID code	Robot ID code	X axis of start position	Y axis of start position
5	6		7	8	9
X axis of target position	Y axis of target position		Orientation	No use	Checksum

Table 1. The communication protocol of the supervised computer.

Byte	0	1	2	3	4
Definition	Start byte	ID code	Robot ID code	X axis of robot	Y axis of robot
5	6		7	8	9
Obstacle status	Orientation of robot		No use	No use	Checksum

Table 2. The communication protocol of the mobile robot.

3. Mobile robot

We develop two module based mobile robot (MBR-I and MBR-II) for the Chinese chess game. The mobile robot (module based robot-I, MBR-I) has the shape of cylinder, and it's equipped with a microchip (MCS-51) as the main controller, two DC motors and driver circuits, a reflect IR sensor circuits, a voice module, an encoder module, three Li batteries and some wireless RF modules. Meanwhile, the mobile robot has four wheels to provide the capability of autonomous mobility.

The mobile robot has some hardware circuits to be shown in Fig. 3. The encoder module uses two reflect IR sensors to calculate the pulse signals from the two wheels of the mobile robot, and calculates the moving distance on the chessboard platform. The power of the mobile robot is three Li batteries, and connects with parallel arrangement. The driver circuits can control two DC motors to execute the displacement command through PWM signal from the controller. The controller of the mobile robot can acquires the detection signals from sensors through I/O pins, and receives the control command via wireless RF interface. The switch input can turns on the power of the mobile robot, and selects power input to be Li batteries or adapter. The voice module can speak the real-time status of the environment for mobile robot moving on the chessboard. In the encoder module, we plot the white line and black line on the wheel of the mobile robot, and use two reflect IR sensors to calculate the pulse signals from the two wheels of the mobile robot. We can set the pulse number for per revolution to be P, and the mobile robot moves pulse number to be B. We can calculate the movement displacement D of the mobile robot using the equation.

$$D = 4.25 \times \pi \frac{B}{P} \tag{1}$$

The diameter of the wheel is 4.25 cm. We design the new chessboard using grid based platform to be shown in Fig. 4. The arrangement of the chess board is 11 grids on the horizontal direction (X axis), and is 12 grids on the vertical direction (Y axis). The distance is 30cm between the center of corridor on the X axis and Y axis of the chessboard. The width of the corridor is 12cm. The mobile robot uses IR sensors to detect obstacles, and decides the cross points of the chess board. The game only uses 9 grids (horizontal) by 10 grids (vertical). We release two grids on each direction (horizontal and vertical) to arrange the leaving chesses of red side and black side. We put the mobile robots that are eaten moving around the platform. We improve the mobile robot (MBR-I) to be implemented in the new chessboard platform, and design new driver device using DC servomotor.

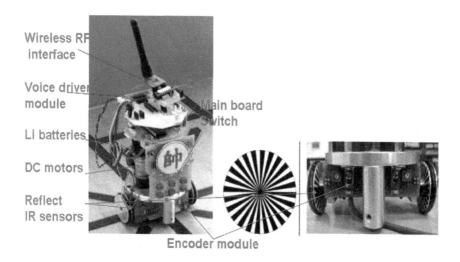

Fig. 3. The structure of the MBR-I.

Fig. 4. The chessboard of the Chinese chess game.

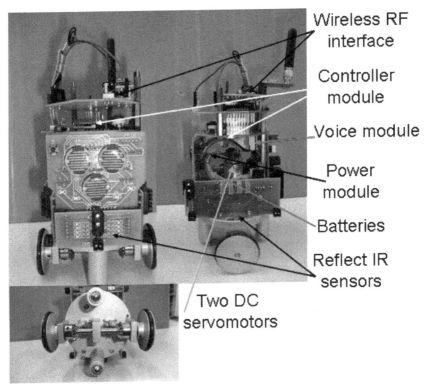

Fig. 5. The structure of the MBR-II.

The mobile robot (module based robot-II, MBR-II) has the shape of cylinder to be equipped with a microchip (MCS-51) as the main controller, and contains two DC servomotors and driver devices, some sensor circuits, a voice module, a compass module, three Li batteries, a wireless RF interface and three reflect IR sensor modules. Meanwhile, the mobile robot has four wheels to provide the capability of autonomous mobility. The structure of the mobile robot is shown in Fig. 5.

The controller calculates the orientation of the mobile robot from the compass module. The compass module has the error range to be 5^0, and transmits the measured values $(x$ and $y)$ in X axis and Y axis, and transmits to the controller of the mobile robot. The mobile robot can calculates the orientation angle θ is

$$\theta = Tan^{-1}\left(-y / x\right) \tag{2}$$

The distance is 30cm between the center of corridor on the X axis and Y axis, and the width of the corridor is 12cm. The mobile robot uses three IR sensors to detect the signals on the front side, right side and left side of the mobile robot, and decides the position of the obstacles, and detect the cross points. The definition of the orientation is east (X- axis), west (X+ Axis), south (Y+ axis) and north (Y- axis) on the motion platform. We set measurement range on the orientation of the mobile robot. For example, the measurement value of the orientation between 226^0 and 315^0, and define the direction of the mobile robot is west.

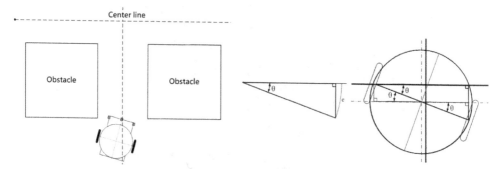

Fig. 6. The mobile robot tuning method.

The mobile robot moves on the corridor of the motion platform, and detects the location to be leaving the center line using the three reflect IR sensors. It tunes the motion path moving to the center line of the corridor. For example, the mobile robot leaves the center line, and moves on the left side of the center line. The mobile robot turns right θ (degree) to be set by users, and is calculated by the encoder module. Then it moves the displacement S according to Eq. (3), and moves to the center line. Finally, the mobile robot turns left θ to face the center line. The overview of the tuning schedule is shown in Fig. 6.

$$S = \frac{R\pi\theta}{180} \tag{3}$$

The parameter R (=4.25cm) is the diameter of the wheel. Furthermore, the mobile robot moves on the right side of the centre line. The tuning processing is the same as the previous method.

4. User interface

The operation interface of the multiple robot based Chinese chess game system is shown in Fig. 7. There are two regions in the operation interface. The chessboard of the Chinese chess is the main monitor of the user interface. Next we explain the chessboard of the Chinese chess game, and describe how to use the interface. It can displays "communication port", "communication protocol" and "axis position" on the bottom side of the operation interface. We set the communication port is 2, and Baud rate is 9600. The start position and target position of each chess displays on the left-bottom side of the operation interface. We make an example to explain the operation interface to be shown in Fig. 7. Players can move "black elephant" to the target position (5,8) from the start position (3,10). The bottom side of the interface can displays the start position (3,10) and the target position (5,8). We develop the operation interface using Visual Basic language for the Chinese chess game system.

We make other example to explain the Chinese chess game interface to be shown in Fig. 8. We can move the chess "red horse" from the start position (2,1) to the next position (3,3) using the mouse. The supervised computer control "red horse" chess moving to the next position (3,3). The scenario of the operation interface is shown in Fig. 8. The bottom side of the operation interface can displays the start position (2,1) and the next position (3,3).

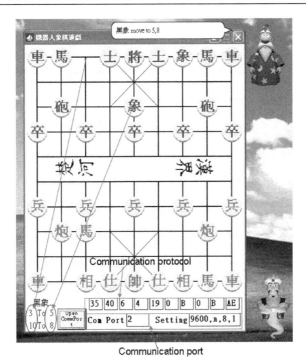

Fig. 7. The operation interface of moving the chess "black elephant".

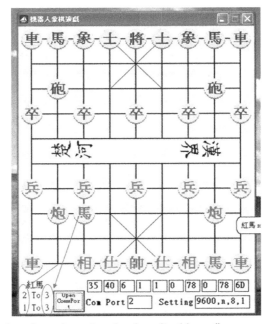

Fig. 8. The operation interface of moving the chess "red horse".

The user interface of the multiple mobile robot based Chinese chess game system has four parts to be shown in left side of the Fig. 9. Players can start or close the game in the part "I". The part "II" can displays the communication protocol on the bottom of the monitor, and the communication port is 2, and Baud rate setting is 9600. The supervised computer transmits the control command to the mobile robot, and receives the status of the mobile robot, and displays the position of the mobile robot moving on the chessboard platform simultaneously to be shown in the part "III".

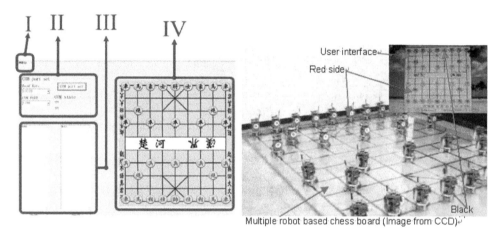

Fig. 9. The user interface of the game.

The part "IV" is chessboard platform. It can displays the motion path that is computed by evaluation algorithm and A* searching algorithm. The operation interface of the multiple robot based Chinese chess game system is shown in right side of the Fig. 9. There are two regions in the user interface. One is real-time image from the image system, and is equipped in the mobile robot based chessboard. The other is the chessboard of the Chinese chess game, and displays on the right side or the monitor. The players can move the chess using the mouse according the rules of the Chinese chess game. Mobile robots receive the status from the supervised computer via wireless RF interface, and move to the next position.

5. Evaluation method

The evaluation algorithm of the Chinese chess game uses the rule based method. Players can move the chess to the target position on the user interface. The position of the chess is not abided by the proposed rules. The user interface can't execute the command, and can't control the mobile robot moving to the target position. Players must renew the operation on the step, and obey the exact rules of the Chinese chess game. We can define the initial position of the chess piece is (x, y), and define the movement rules of all chesses as following. n is movement displacement on the x axis, and m is movement displacement on the y axis. n and m of the target position must be plus integer. The positions of all chesses must obey the rules that are listed in the Table 3 at any time.

Chess	Target position	Rules
Red king	$(x\pm n,y\pm m),\ 0\le n\le 1,\ 0\le m\le 1$	$3\le x\le 5,\ 3\le x\pm n\le 5,$ $0\le y\le 2,\ 0\le y\pm m\le 2$
Black king	$(x\pm n,y\pm m),\ 0\le n\le 1,\ 0\le m\le 1$	$3\le x\le 5,\ 0\le x\pm n\le 5,$ $7\le y\le 9,\ 0\le y\pm m\le 9$
Red advisor	$(x\pm n,y\pm n),\ n=1$	$3\le x\le 5,\ 0\le x\pm n\le 5,$ $0\le y\le 2,\ 0\le y\pm n\le 2$
Black advisor	$(x\pm n,y\pm n),\ n=1$	$3\le x\le 5,\ 0\le x\pm n\le 5,$ $7\le y\le 9,\ 0\le y\pm n\le 9$
Red elephant	$(x\pm n,y\pm n),\ n=2$	$0\le x\le 8,\ 0\le x\pm n\le 8,$ $0\le y\le 4,\ 0\le y\pm n\le 4$
Black elephant	$(x\pm n,y\pm n),\ n=2$	$0\le x,x\pm n\le 8,\ 5\le y,y\pm n\le 9$
Red rook, Red cannon Black rook, Black cannon	$(x\pm n,y\pm m),\ n=0\ \text{or}\ m=0$	$0\le x,x\pm n\le 8,\ 5\le y,y\pm m\le 9$
Red horse, Black horse	$(x\pm n,y\pm m),\ n+m=3$	$0\le x\le 8,\ 0\le x\pm n\le 8,$ $0\le y\le 9,\ 0\le y\pm m\le 9$
Red pawn	$(x,y+1)$	$0\le x\le 8,$ $3\le y\le 4,\ 3\le y\pm m\le 4$
	$(x\pm n,y+m),\ n+m=1$	$0\le x\le 8,\ 0\le x\pm n\le 8,$ $5\le y\le 9,\ 5\le y+m\le 9$
Black pawn	$(x,y-1)$	$0\le x\le 8,$ $5\le y\le 6,\ 5\le y\pm m\le 6$
	$(x\pm n,y-m),\ n+m=1$	$0\le x\le 8,\ x\pm n\le 8,$ $0\le y\le 4,\ 0\le y-m\le 4$

Table 3. The rules of Chinese chesses.

Mobile robots move on the chessboard to obey the rules of Chinese chess game. The user interface knows the start position and the target position, and programs the shortest path of the chess (mobile robot) moving to the target position from the start position. We use A* searching algorithm to find the shortest path on the Chinese chess game. A* searching algorithm is proposed by Hart in 1968, and solved the shortest path problem of multiple nodes travel system. The formula of A* searching algorithm is following

$$f(n)=g(n)+h(n) \tag{4}$$

The core part of an intelligent searching algorithm is the definition of a proper heuristic function $f(n)$. $g(n)$ is the exact cost at sample time n from start point to the target point. $h(n)$ is an estimate of the minimum cost from the start point to the target point. In this study, n is reschedules as n' to generate an approximate minimum cost schedule for the next point. The equation (5) can be rewritten as follows:

$$f(n) = g(n) + h(n') \tag{5}$$

We make an example to explain algorithm. Such as a mobile robot move to the target point "T" from the start point "S". The position of the start point is (2,6), and the target position is (2,9). We set some obstacle on the platform. The white rectangle is unknown obstacle. The black rectangle (obstacle) is detected by the mobile robot using A* searching algorithm. We construct two labels (Open list and Close list) in the right side of Fig. 8. The neighbour points of the start point fill in the "Open list". The "Close list" fills the start point and evaluation points. We construct label on the first searching result to be shown in Fig. 8. We calculate the values of $f(n)$, $g(n)$ and $h(n)$ function according to the pulse numbers by the encoder of the DC servomotor, and use the proposed method to compare the values of the function. We select the minimum value of the function $f(n)$ to be stored in "Close list". We can find the target point on the final searching result to be shown in Fig. 10, and we can decide a shortest path to control the mobile robot moving to the target point.

The total distance of the shortest path C_{st} can be calculated as

$$C_{st} = \sum_{n=1}^{m} G_n (m = t - 1) = 2043 \tag{6}$$

In the other condition, we rebuild the positions of the obstacles on the platform to be shown in Fig. 11. The mobile robot can't find the path C_{st} moving to the target position using the proposed method. The total distance C_{st} as

$$C_{st} = \infty \tag{7}$$

Open	Closed	Point	F_{12}	G_{12}	H_{12}	D
(3,6)	S	(3,6)	1380	732	648	→
(1,6)	(2,7)	(1,6)	1380	732	648	↑
(1,7)	(3,7)	(1,7)	1056	570	486	↑
(6,8)	(3,8)	(6,8)	1704	864	810	↑
(6,9)	(4,8)	(6,9)	1338	690	648	↑
(6,10)	(5,8)	(6,10)	1662	852	810	↑
(4,11)	(5,9)	(4,11)	1338	690	648	↑
(3,11)	(5,10)	(3,11)	1014	528	486	↑
(2,11)	(4,10)	(2,11)	648	162	324	↑
	(3,10)					
	(2,10)					
	T					

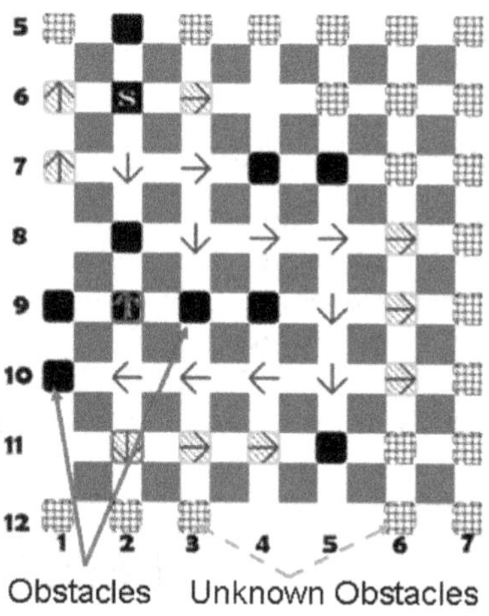

Fig. 10. The searching result for case 1.

Obstacles   Unknown Obstacles

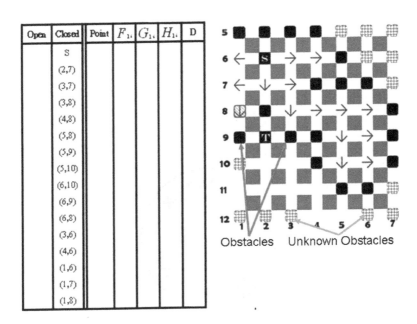

Open	Closed	Point	F_{1i}	G_{1i}	H_{1i}	D
	S					
	(2,7)					
	(3,7)					
	(3,8)					
	(4,8)					
	(5,8)					
	(5,9)					
	(5,10)					
	(6,10)					
	(6,9)					
	(6,8)					
	(3,6)					
	(4,6)					
	(1,6)					
	(1,7)					
	(1,8)					

Obstacles   Unknown Obstacles

Fig. 11. The searching result for case 2.

6. Experimental results

We execute some experimental scenarios using the MBR-I robot on the chessboard for the Chinese chess game system, and test the movement functions of some chesses (mobile robots). The two players are located by two sides, and move the chess one time each other. One moves red chesses; the other moves black chesses. The first experimental scenario is "red king". The player moves forward the chess "red king" using the mouse to be shown in Fig. 12 (a). The chess "red king" moves one grid to the target position on the user interface to be shown in Fig. 12 (b). The supervised computer orders the command to the mobile robot "red king" moving forward to the target position via wireless RF interface. The mobile robot can calculates the movement displacement using the encoder module, and speaks the movement status using the voice module. The experimental result is shown in Fig. 12 (c).

The second experimental scenario is "red rook". The player moves forward chess "red rook" using the mouse to be shown in Fig. 13 (a). The supervised computer orders the command

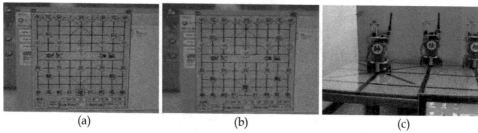

Fig. 12. The experimental result for "red king".

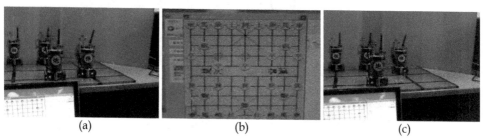

Fig. 13. The experimental result for "red rook".

to the mobile robot "red rook". The mobile robot "red rook" moves to the target position (8,2) from the start position (8,0). The mobile robot can calculates the movement displacement using the encoder module, and stops at the target position. The experimental result is shown in Fig. 13 (b) and (c).erwe

The third experimental scenario is "red cannon". The player moves the chess "red cannon" to right side using the mouse. The chess "red cannon" moves to the target position (4,2) from the start position (1,2) to be shown in Fig. 14 (a) and (b). The supervised computer orders the command to the mobile robot "red cannon" moving to the target position. The mobile robot receives the command to turn right, and moves to the target position. Then it can turn left and face to the black chess. The mobile robot calculates the movement displacement and the orientation using the encoder module, and speaks the movement status using the voice module. The experimental results are shown in Fig. 14 (c) and (d).

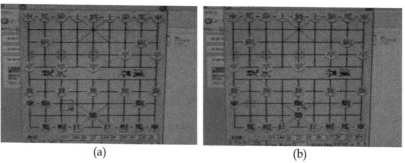

Fig. 14. (Continued)

(c)   (d)

Fig. 14. The experimental result for "red cannon".

We implement the functions of the mobile robot to obey the movement rules of Chinese chess game. Then we execute experimental scenarios of the multiple mobile robots based Chinese chess game system. The first experimental scenario is "red elephant" that is moved by the player (red side). The player moves the chess "red elephant" to the target position (4,2) using the mouse to be shown in Fig. 15 (a) and (b). The supervised computer orders the command to the mobile robot "red elephant" via wireless RF interface. The start position of the mobile robot is (6,0). The movement orientation of the mobile robot is west-north. The mobile robot receives the command to turn left on 90⁰, and moves forward to the target position according to the Chinese chess rules. The mobile robot speaks the movement status using voice module, and calculates the movement displacement using the encoder module, and moves to the target position (4,2). The experimental results are shown in Fig. 15 (c) and (d).

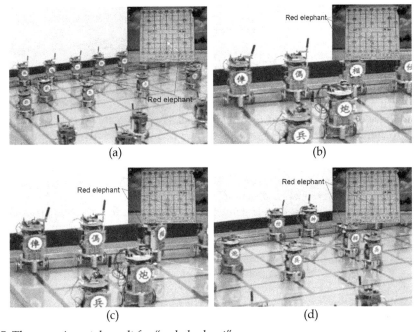

Fig. 15. The experimental result for "red elephant".

The second experimental scenario is "black advisor" that is moved by the other player (black side). The player moves the chess "black advisor" using the mouse to be shown in Fig. 16 (a). The start position of the "black advisor" chess is (3,9). The supervised computer orders the command to the mobile robot "black advisor". The chess turn left on 90⁰, and moves to the target position (4,8) and stop. The mobile robot can calculates the movement displacement using the encoder module, and speak the movement status using voice module, too. The experimental result is shown in Fig. 16 (b). The next step must moves red chess by the player (red side). The players (red side and black side) move one chess step by step on the chessboard each other. Finally, the user interface can decides which player to be a winner, and closes the game.

(a)   (b)

Fig. 16. The experimental result for "black advisor".

The controller of the mobile robots calculates the movement displacement to be error, or loses pulse numbers from the encoder module. Mobile robots can't stay at the cross point of two black lines on the chessboard for a long time. We design the grid based chessboard platform to improve the weakness, and implement some experimental results using the multiple mobile robots (MBR-II) based Chinese chess game system. The first experimental scenario is "red elephant".

The user moves the chess "red elephant" using the mouse to be shown in Fig. 17 (a). The start position of the "red elephant" chess is (6,0). The supervised computer orders the command to the mobile robot "red elephant". The chess moves forward two grids, and turn left angle to be 90⁰. Then the mobile robot moves two grids to the next position (4,2), and turn right angle 90⁰ to face the black side and stop. The user interface of the Chinese chess game system uses multiple mobile robots to be shown in Fig. 17(a). The mobile robot can calculates the movement displacement via encoder of DC servomotor, and speaks the movement status of the mobile robot using voice module, too. The experimental scenarios use the mobile robot to execute the motion path of the chess "red elephant" to be shown in Fig. 17 (b)-(e). There is not obstacle on the motion path of the chess. We program the shortest path using A* searching algorithm.

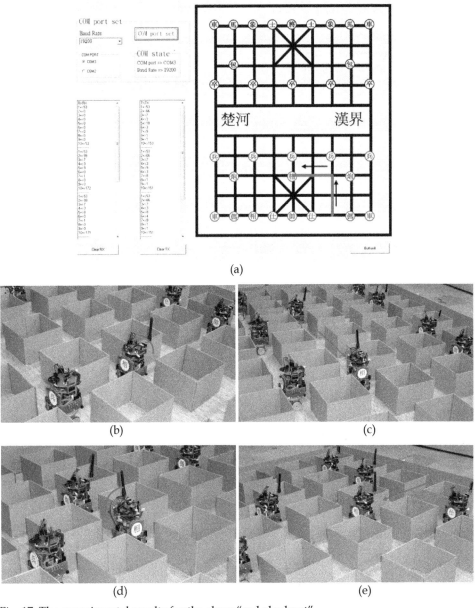

(a)

(b)   (c)

(d)   (e)

Fig. 17. The experimental results for the chess "red elephant".

The second experimental scenario is "red cannon". The user moves the chess "red cannon" using the mouse to be shown in Fig. 18 (a). The start position of the "red cannon" chess is (1,2), and the target point is (1,6). The mobile robot "red cannon" can't moves forward to the target position (6,1). The motion path of the mobile robot has an obstacle (chess) at the position (1,5). The supervised computer reconstructs all possible paths using A* searching algorithm for red cannon step by step. Finally, we can find out the shortest path to avoid the obstacle (chess). The path can displays on the interface using red line. The supervised computer controls the mobile robot moving to the target point from the start point using the shortest path to avoid the obstacle. The supervised computer calculates the cost $f(n)$ to be listed on the left side of the Fig. 18 (a).

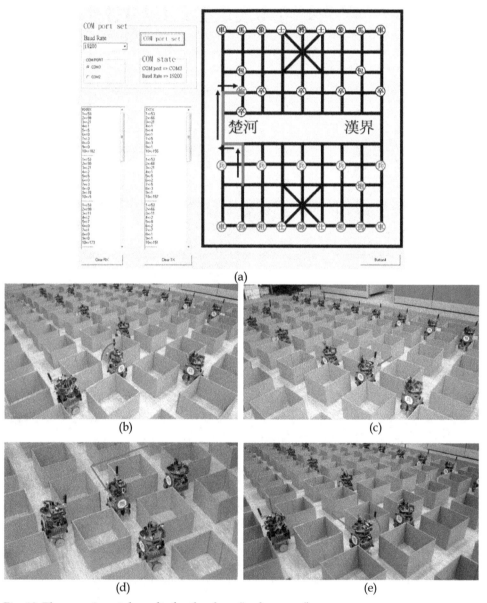

(a)

(b)   (c)

(d)   (e)

Fig. 18. The experimental results for the chess "red cannon".

The supervised computer orders the command to the mobile robot "red cannon". The chess piece moves forward two grids, and turn left angle to be 90^0. Then the mobile robot moves one grid to, and turn right angle 90^0. Then the mobile robot moves two grids, and turn right angle 90^0. Finally the mobile robot moves one grid to the target position (1,6), and turn left angle 90^0 to face the black side and stop. Finally, the experiment scenarios are shown in Fig. 18(b)-(e).

7. Conclusion

We have presented the experimental scenario of the Chinese chess game system using multiple mobile robots. The system contains a supervised computer, a image system, some wireless RF modules and thirty-two mobile robots. Mobile robots are classified red side and black side. We design two types' mobile robots (MBR-I and MBR-II) for Chinese chess game. The two mobile robots have the shape of cylinder and its diameter, height and weights is 8cm, 15cm and 1.5kg, and execute the chess attribute using two interfaces. One is wireless RF interface, and the other is voice interface. We develop the user interface on the supervised computer for the Chinese chess game. The supervised computer can programs the motion paths of the mobile robots using evaluation algorithm according to the rule of Chinese chess game, and receive the status of mobile robots via wireless RF interface. The MBR-I calculates the movement displacement using the pulse numbers from reflect IR sensors of the encoder module. Then we develop the MBR-II robot to implement scenario of the Chinese chess game on the grid based chessboard platform. The MBR-II robot calculates the movement displacement using encoder of DC servomotor. The chess (mobile robot) uses A* searching algorithm to program the shortest path moving to the target point. The supervised computer can controls mobile robots, and receives the status of mobile robots via wireless RF interface. Players can move the chess piece using the mouse on the supervised computer, and obey the game rules. The supervised computer can controls the mobile robot moving to the target position. The mobile robot speaks Chinese language for the movement status.

8. Acknowledgment

This work was supported by the project "Development of an education robot", under National Science Council of Taiwan, (NSC 99-2622-E-224-012-CC3).

9. References

S. J. Yen, J. C. Chen, T. N. Yang and S. C. Hsu, "Computer Chinese Chess," ICGA Journal, Vol.27, No. 1, pp.3-18, Mar, 2004.

P. Darwen and X. Yao, "Coevolution in Iterated Prisoner's Dilemma with Intermediate Levels of Cooperation: Application to Missile Defense," International Journal of Computational Intelligence Applications, Vol. 2, No. 1, pp.83-107, 2002.

C. H. Yong and R. Miikkulainen, "Cooperative Coevolution of Multi-agent Systems," University of Texas, Austin, USA, Tech. Rep. AI01-287,2001.

C. S. Ong, H. Y. Quek, K. C. Tan and A. Tay, "Discovering Chinese Chess Strategies Through Coevolutionary Approaches," IEEE Symposium on Computational Intelligent and Games, pp.360-367, 2007.

J. Wang, Y. H. Luo, D. N. Qiu and X. H. Xu, "Adaptive Genetic Algorithm's Implement on Evaluation Function in Computer Chinese Chess," Proceeding of ISCIT 2005, pp.1206-1209.

W. P. Lee, L. J. Liu and J. A. Chiou, "A Component-Based Framework to Rapidly Prototype Online Chess Game for Home Entertainment," IEEE International Conference on System, Man and Cybernetics, pp.4011-4016, 2006.

W. Zhou, J. L. Zhang and Y. Z. Wang, "The Iterative Sort Search Techniques Based on Percentage Evaluation," Chinese Control and Decision Conference (CCDC 2008), pp.5263-5266.

Kuo-Lan Su, Sheng-Ven Shiau, Jr-Hom Guo and Chih-Wei Shiau, "Mobile Robot Based Onlin Chinese Chess Game," The Fourth International Conference on Innovative Computing, Information and Control, pp.63, 2009.

C. Buiu and N. Popescu, Aesthetic emotions in human-robot interaction implications on interaction design of robotic artists, International Journal of Innovative Computing, Information and Control, Vol.7, No. 3, pp.1097-1107, 2011.

T. Song, X. Yan, A Liang and K. Chen, A distributed bidirectional auction algorithm for multi-robot coordination, IEEE International Conference on Research Challenges in Computer Science, pp.145-148, 2009.

R. Grabowski and L. Navarro-Serment, C. Paredis, and P. Khosla, Heterogeneous teams of modular robots for mapping and exploration, Autonomous Robots, Vol.8, No. 3, pp.293-308, 2000.

E. J. Chung, Y. S. Kwon, J. T. Seo, J. J. Jeon, andH. Y. Lee, Development of a multiple mobile robotic system for team work, SICE-ICASE International Joint Conference, pp.4291-4296, 2006.

T. Balch, and R. Arkin, Behavior-based formation control for multirobot teams, IEEE Trans. on Robotics and Automation, Vol.14, No. 6, pp.926-939, 1998.

R. Alami, S. Fleury, M. Herrb, F. Ingrand, and F. Robert, Multi-robot cooperation in the MARTHA project, IEEE Robotics and Automation Magazine, Vol. 5, No. 1, pp.36-47, 1998.

Y. Cao et al, Cooperative mobile robotics: antecedents and directions, Autonomous Robots, Vol.4, No. 1, pp.7-27, 1997.

J. H. Guo and K. L. Su, Ant system based multi-robot path planning, ICIC Express Letters, Part B: Applications, Vol. 2, No. 2, pp. 493-498, 2011.

W. Burgard, M. Moors et al, Coordinated Multi-Robot Exploration, IEEE Transaction on Robotics, Vol. 21, No. 3, pp.376-386, 2005.

K. L. Su, C. Y. Chung, Y. L. Liuao and J. H. Guo, A* searching algorithm based path planning of mobile robots, ICIC Express Letters, Part B: Applications, Vol. 2, No. 1, pp. 273-278, 2011.

Y. Saber and T. Senjyu (2007) Memory-bounded ant colony optimization with dynamic programming and A* local search for generator planning, IEEE Trans. on Power System, Vol.22, No. 4, pp.1965-1973

Permissions

The contributors of this book come from diverse backgrounds, making this book a truly international effort. This book will bring forth new frontiers with its revolutionizing research information and detailed analysis of the nascent developments around the world.

We would like to thank Andon V. Topalov, for lending his expertise to make the book truly unique. He has played a crucial role in the development of this book. Without his invaluable contribution this book wouldn't have been possible. He has made vital efforts to compile up to date information on the varied aspects of this subject to make this book a valuable addition to the collection of many professionals and students.

This book was conceptualized with the vision of imparting up-to-date information and advanced data in this field. To ensure the same, a matchless editorial board was set up. Every individual on the board went through rigorous rounds of assessment to prove their worth. After which they invested a large part of their time researching and compiling the most relevant data for our readers. Conferences and sessions were held from time to time between the editorial board and the contributing authors to present the data in the most comprehensible form. The editorial team has worked tirelessly to provide valuable and valid information to help people across the globe.

Every chapter published in this book has been scrutinized by our experts. Their significance has been extensively debated. The topics covered herein carry significant findings which will fuel the growth of the discipline. They may even be implemented as practical applications or may be referred to as a beginning point for another development. Chapters in this book were first published by InTech; hereby published with permission under the Creative Commons Attribution License or equivalent.

The editorial board has been involved in producing this book since its inception. They have spent rigorous hours researching and exploring the diverse topics which have resulted in the successful publishing of this book. They have passed on their knowledge of decades through this book. To expedite this challenging task, the publisher supported the team at every step. A small team of assistant editors was also appointed to further simplify the editing procedure and attain best results for the readers.

Our editorial team has been hand-picked from every corner of the world. Their multi-ethnicity adds dynamic inputs to the discussions which result in innovative outcomes. These outcomes are then further discussed with the researchers and contributors who give their valuable feedback and opinion regarding the same. The feedback is then collaborated with the researches and they are edited in a comprehensive manner to aid the understanding of the subject.

Apart from the editorial board, the designing team has also invested a significant amount of their time in understanding the subject and creating the most relevant covers. They scrutinized every image to scout for the most suitable representation of the subject and create an appropriate cover for the book.

The publishing team has been involved in this book since its early stages. They were actively engaged in every process, be it collecting the data, connecting with the contributors or procuring relevant information. The team has been an ardent support to the editorial, designing and production team. Their endless efforts to recruit the best for this project, has resulted in the accomplishment of this book. They are a veteran in the field of academics and their pool of knowledge is as vast as their experience in printing. Their expertise and guidance has proved useful at every step. Their uncompromising quality standards have made this book an exceptional effort. Their encouragement from time to time has been an inspiration for everyone.

The publisher and the editorial board hope that this book will prove to be a valuable piece of knowledge for researchers, students, practitioners and scholars across the globe.

List of Contributors

Farouk Azizi and Nasser Houshangi
Purdue University Calumet, USA

Yuanlong Yu, George K. I. Mann and Raymond G. Gosine
Faculty of Engineering and Applied Science, Memorial University of Newfoundland, St. John's, NL, Canada

D.-L. Almanza-Ojeda and M.-A. Ibarra-Manzano
Digital Signal Processing Laboratory, Electronics Department; DICIS, University of Guanajuato, Salamanca, Guanajuato, Mexico

Ryosuke Kawanishi, Atsushi Yamashita and Toru Kaneko
Shizuoka University, Japan

Mehmet Serdar Guzel and Robert Bicker
Newcastle University, United Kingdom

Junghee Park and Jeong S. Choi
Korea Military Academy, Korea

Liying Yang
State Key Laboratory of Robotics, Shenyang Institute of Automation, Chinese Academy of Sciences, Shenyang, China
Graduate School, Chinese Academy of Sciences, Beijing, China

Jianda Han
State Key Laboratory of Robotics, Shenyang Institute of Automation, Chinese Academy of Sciences, Shenyang, China

Yang Chen
State Key Laboratory of Robotics, Shenyang Institute of Automation, Chinese Academy of Sciences, Shenyang, China
Department of Information Science and Engineering, Wuhan University of Science and Technology, Wuhan, China
Graduate School, Chinese Academy of Sciences, Beijing, China

Valeri Kroumov
Department of Electrical & Electronic Engineering, Okayama University of Science, Okayama, Japan

Jianli Yu
Department of Electronics and Information, Zhongyuan University of Technology, Zhengzhou, China

Andrey Fedotov and Valerii Patsko
Institute of Mathematics and Mechanics, Ural Branch of the Russian Academy of Sciences, Russia

Varvara Turova
Mathematics Centre, Technische Universität München, Germany

Charmane V. Caldwell, Emmanuel G. Collins and Jr. and Oscar Chuy
Center for Intelligent Systems, Control and Robotics (CISCOR), FAMU-FSU College of Engineering, USA

Damion D. Dunlap
Naval Surface Warfare Center - Panama City Division, USA

Jr-Hung Guo and Sheng-Ven Shiau
Graduate school Engineering Science and technology, National Yunlin University of Science & Technology, Taiwan

Kuo-Lan Su
Department of Electrical Engineering, National Yunlin University of Science & Technology, Taiwan